KW-020-227

IUCN
Social Policy Group

Population and Strategies for National Sustainable Development

A Guide to Assist National Policy Makers in
Linking Population and Environment in
Strategies for Sustainable Development

Gayl D. Ness with Meghan V. Golay

Earthscan Publications Ltd, London

Population and Strategies for National Sustainable Development: A Guide to Assist National Policy Makers in Linking Population and Environment in Strategies for Sustainable Development was made possible by the generous support of the United Nations Population Fund (UNFPA).

First published in the UK in 1997 by
Earthscan Publications Limited

Copyright © International Union for Conservation of Nature and Natural Resources, 1997

All rights reserved

A catalogue record for this book is available from the British Library

ISBN: 1 85383 375 4

Typesetting by PCS Mapping & DTP, Newcastle upon Tyne

Page design by S&W Design

Printed and bound by Biddles Ltd, Guildford and King's Lynn

Cover design by Elaine Marriott

Cover photograph by Grazia Borrini-Feyerabend

The views of the authors expressed in this book do not necessarily reflect those of IUCN. The designations of geographical entities in this book, and the presentation of the material, do not imply the expression of any opinion whatsoever on the part of IUCN concerning the legal status of any country, territory or area, or of its boundaries.

For a full list of publications please contact:

Earthscan Publications Limited
120 Pentonville Road
London N1 9JN
Tel: (0171) 278 0433
Fax: (0171) 278 1142
Email: earthinfo@earthscan.co.uk

or visit our Web Site at

http://www.earthscan.co.uk

Earthscan is an editorially independent subsidiary of Kogan Page Limited and publishes in association with the WWF-UK and the International Institute for Environment and Development.

Printed on acid and elemental chlorine free paper, sourced from sustainably managed forests and processed according to an environmentally responsible manufacturing system.

Contents

Acknowledgements

We have received much help and encouragement in producing this volume. Many IUCN and UNFPA field and headquarters personnel offered comments on an earlier draft, and many more contributed to our understanding of the issues and the problems. While we cannot acknowledge all the debts, a few must be noted for their special role or for the extent and detail of their comments and specific suggestions. Jeremy Carew Reid and Grazia Borrini provided continued encouragement and critical discussion of many of the fundamental issues we attempt to address here. Mersie Ejigu and Per Ryden of IUCN headquarters provided important suggestions that have shaped the overall structure of the Guide. For detailed comments, special appreciation must go to Dr. Paul Harrison, Dr. George Martine, Dr. Ralph Hakkert, and Alex de Sherbinin. Dr. Serge Ivanov of the UN Population Division provided useful comments, as did Arnefin Jorgensen-Dahl, Satish Merah, Abdel Kader Fahem and Zulma Recchini de Lattes of UNFPA. In any such work as this, there are many differences of interpretation and emphasis. We hope it is evident that we have taken seriously all of the comments, even when we have not accepted them. We can only hope that those who provided comments will see here our attempt to deal fairly with them, even when we disagree.

The IUCN provided a general setting that was rich in experience in environmental conservation, which emerged in informal discussions over lunch, in the halls and in the field. Three important outside institutional supports must also be identified. The University of Michigan's USAID funded Population Fellows Program provided the senior author two exceptionally rewarding years to work with the IUCN in Gland. The UNFPA provided funds for assistance in preparing this volume, and for its companion, *Our People, Our Resources*, which is directed at community level integration of population and environmental issues. USAID, UNFPA and the University of Michigan are to be commended for the vision their leaders display in establishing institutions and processes that support attempts to deal with economic and human development issues in a broad and humane manner.

We alone, however, take responsibility for the views here. They are those of the authors, and do not necessarily represent the views of the UNFPA, or of IUCN, its members and partners.

Executive Summary

The best way to summarize this book is to raise a series of critical questions, and direct the reader to responses in the text. There are many critical questions in attempting to link population with strategies for sustainable development. The following are 12 we find most important and relevant.

1. WHY TRY TO LINK POPULATION WITH ENVIRONMENTAL ISSUES IN NATIONAL STRATEGIES?

Page vii ff. The Preface makes some basic arguments, and identifies bases for action in a variety of global agreements coming out of international conferences.

Page 21 ff. Provides more reasons: population and environment are linked in the real world, and both can be treated best in connection with the other.

Page 59 ff. Specifies some of the substantive questions found in common frameworks and simulation models of population, development and environment.

2. WHAT IS A NATIONAL STRATEGY, AND WHAT KIND OF DEVELOPMENT IS IT AIMING AT?

Page 15 ff. Identifies National Strategies for Sustainable Development, and specifies the meaning of development and of sustainable development:
Promoting the wellbeing of people and ecosystems.

3. WHY IS IT DIFFICULT TO MAKE THE LINKAGES?

Page 21 ff. Identifies the problem in the strength of specialization, and suggests that what is needed is bridges across specializations. These are primarily organizational problems; we can turn to experience for lessons.

4. WHAT ABOUT INTEGRATION?

Page 30 ff. Probably the most used and abused term in this business. It seems everyone wants to do it, and, even more, everyone wants the magic formula for how to do it. There is none. We can offer some ideas, but it must always be done by adapting to local conditions.

5. ANY GOOD IDEAS?

Page 40 ff. Possibly. Here we propose a strategy for building Population Environment Networks, or PENs.

6. WHAT'S ALL THIS ABOUT I = PAT?

Page 57 ff. Introduces frameworks, or intellectual tools we use for thinking about population environment issues. I = PAT is one of the most common frameworks, and has been used in a variety of ways to help us think about the problem.

7. AND MODELS? WHY MODELS? WHAT MODELS?

Page 70 ff.   Introduces the more complex business of building quantitative models to help us understand how things work, and to try to look into the future to see the probable outcomes of current actions, or of different policy options.

8. WHAT ABOUT THE DEMOGRAPHIC TRANSITION?

Page 89 ff.   One of the more important observations to be made in the history of population change. There are not one but two: the past and the present. Here we make the case that the current form of the demographic transition indicates that we now have far more control over human mortality and fertility than at any time in the past.

9. POPULATION PROBLEMS – WHAT CONDITIONS ARE IMPORTANT?

Page 99 ff.   Introduces a large number of basic conditions of the human population, and how we can measure those conditions. They come out of the specialization of demography or population studies, and are the things we need to pay attention to in planning for sustainable development.

10. WHAT ABOUT MORTALITY AND FERTILITY?

Page 103 ff.   Both must be reduced in the less developed regions to promote the wellbeing of people and ecosystems. Today it is far easier to do than in the past, and governments have a major responsibility for doing it.

11. AND MIGRATION?

Page 118 ff.   An inexorable process, as old as the human species itself, it is both caused by and produces environmental change. Few governments have been able to control migration, but it can be predicted and at least in part planned for.

12. AND URBANIZATION?

Page 122 ff.   It appears inevitable, and it may well be the best strategy for promoting sustainable development. Many governments try to slow the process; it would be better to work to promote sustainable cities.

Preface
Addressing Population–Environment Linkages to Promote Sustainable Development

Linkages for Sustainable Development

In the past three decades the linkage between population and environmental problems has become a major object of concern in international development. The world's population has reached the unprecedented level of 5.7 billion, and though rates of growth have slowed, it is still possible that the population will double or more before levelling off. The optimistic hopes for economic development of the 1960s have faded. Despite large amounts of international assistance, and some successes, the list of problems grows, along with increasing human misery. Environmental degradation reaches all corners of the planet, from ozone destruction, to global warming, to deforestation and life threatening pollution. Too often, however, population, development and environment have been treated separately, following lines of specialization that divide scientific disciplines and development agencies alike.

This separation is now recognized as a major problem, giving rise to new and urgent calls for more attention to linking population and environmental issues in promoting sustainable development, which is itself a new idea that is an adaptation to the failures of the development dream.

The human species, which grew very slowly for thousands of years, suddenly exploded after 1950, with growth rates peaking at over 2 percent per year. Now the large population base means large absolute additions, possibly near 1 billion per decade over the next half century or more. Moreover, the rapid growth will occur in the poorer countries of the world, those least capable of coping with it.

Over the past three decades and more, average annual world economic growth has generally exceeded world population growth by one or two percentage points. Even with the shocks of rising oil prices in the 1970s and the world wide recession of the 1980s, economic growth overall has kept ahead of population growth. In reviewing three decades of work on poverty, the World Bank estimated average annual real per capita growth rates over 1950–80 at 3 percent for industrial and middle income countries, and even 1.3 percent for the low income countries. By these measures economic development, almost universally desired, has been almost universally experienced. But the gap between rich and poor remains, in some cases grows larger, and in all cases is both unstable and unsustainable.

There have been some real gains in the quality of life, at least by major aggregate measures. These gains themselves have led to rapid population growth. Infant and maternal mortality have declined in most areas, life expectancy has risen, along with school enrolment and literacy, especially for the most disadvantaged of the human population, women and girls.

But other gains remain elusive, and the rise of absolute numbers in stark poverty overwhelms whatever gains may be apparent for some. The number of people in absolute poverty rises daily. The numbers of hungry, sick and malnourished people do not diminish. Deaths from simple, easily controlled infectious diseases number in the hundreds of thousands annually.

And whatever achievements we make seem to come at high costs to the environment. Forests are diminishing; air, water and land pollution grow rapidly; biodiversity is declining; the protective ozone layer is being destroyed and the planet is threatened with future temperature rises that could spell disaster for many areas.

The linkages between these dynamics and their attendant problems are complex, but visible everywhere. They are especially evident in the world's unequal distribution of wealth and welfare. Poor people are often driven to farm weak and vulnerable soils because they have nowhere else to go for food, while wealthy farmers turn forests into pastures for beef to earn foreign exchange. Poor women in poor countries often bear more children than they wish because they lack the basic social services that give them a choice in fertility and family size. In many wealthy countries, on the other hand, where social services are extensively available, populations are expected to decline. Wealthy countries can protect their forest resources by promoting destructive logging in poorer countries. Wealthy countries can export toxic wastes and toxic industries to poor countries where environmental protection systems are weak. Environmental organizations in wealthy countries support biodiversity and protected areas in poor countries, often excluding indigenous peoples whose lives depend on the resources being protected, and who have often lived for centuries using those resources in a sustainable fashion. These global inequalities are clearly unstable and unsustainable. They tend to lead to conflicts between environmental conservation and human welfare.

Although the linkages between population, development and environment are complex and often conflicting, it is possible to address the linkages in such a manner that promotes the welfare of both people and the environment. But to do this, we shall have to change common ways of thinking and acting.

The narrow emphasis on economic growth should give way to a focus on sustainable development. The fundamental aspect of sustainable development is that it promotes the welfare of both people and ecosystems, implying increasing human productivity to raise the level of human welfare for both present and future generations.

The narrow emphasis on environmental conservation, often implying exclusion of people, should give way to an emphasis on human sustainable use of natural resources. This implies linking people and population dynamics to environmental conservation in a mutually beneficial system.

The narrow emphasis on population control or fertility limitation should give way to an emphasis on human welfare, reproductive health, responsible parenthood, and choice. This implies providing basic social services, including primary health care, education and family planning, especially to women and to people in rural areas.

Although current emphases and programmes are now broadening and changing in positive directions, much remains to be done to make the linkages between population and sustainable development more productive. For this we need both more knowledge and more action. Both knowledge and action require that we find ways to bridge the scientific and organizational specializations that mitigate against cooperative effort.

Scientific specializations have given us great powers of observation and knowledge generation. From agriculture and anthropology to sociology and zoology, scientific disciplines have greatly increased our knowledge about the world precisely because they have focused observation and analysis on a narrow range of conditions. These have often been paralleled by organizations that specialize in specific activities, such as curing illness, providing health care, promoting agriculture or building effective and efficient water and waste systems.

It is also those specializations, however, that make linking difficult. Specializations work in large part because they build strong agendas and tools for action and observation. But these often act as barriers to communication across specializations. Linking population with sustainable development will require that some barriers be dismantled, or that bridges be built between specialized disciplines.

This guide is concerned with making the linkages and building the bridges between population dynamics and attempts to promote sustainable development, especially at the level of national planning. Another manual published by IUCN, *Our People, Our Resources* (Barton et al, 1996), provides tools and options for making the linkages at the local community level.

Bases for Action

The grounds for action by the global community and by national governments have grown rapidly over the past two decades. The first world conference on the environment was held in Stockholm in 1972. Although it articulated some controversies between the more and less developed regions, it eventually led to the formation of the United Nations Environment Programme (UNEP), and to the United Nations Conference on Environment and Development (UNCED) held in Rio de Janeiro in 1992 (Robinson, 1993).

From UNCED came *Agenda 21*, a 700 page document in which nations of the world agreed that today:

> *Humanity stands at a defining moment in history. We are confronted with a perpetuation of disparities between and within nations, a worsening of poverty, ill-health, and illiteracy and the continuing deterioration of the ecosystem on which we depend for our well-being. However, integration of environment and development concerns, and greater attention to them will lead to fulfilment of basic needs, improved living standards for all, and a safer, more prosperous future. (Agenda 21, Preamble, 1.1)*

In 1969 the United Nations created the Fund for Population Activities (UNFPA) and in 1974 convened the first International Conference on Population, held in Bucharest, Romania. There, too, despite initial controversies, the Conference adopted a World Plan of Action that emphasized an integrated approach to population and development issues. Two subsequent decennial conferences were held in Mexico City in 1984, and Cairo in 1994. The Cairo Conference was officially designated the International Conference on Population and Development. The General Assembly resolution 49/128, 'Report of the International Conference on Population and Development' as adopted on 19 December 1994, among other things:

> *(6.) Fully acknowledges that the factors of population, health, education, poverty, patterns of production and consumption, empowerment of women and the environment are closely interconnected and should be considered through an integrated approach...*

The Cairo Conference was especially important in giving emphasis to the condition of women as a determinant of modern population dynamics. That raised some controversy, but in the end there was broad agreement that better primary health care, more equal social services and opportunities, and better family planning services for women are both morally imperative and necessary to achieve real advances in human welfare. The consensus at Cairo has been endorsed by some 180 countries.

Finally, the World Summit for Social Development took another step in this integrated march of international conferences by focusing on 'Attacking Poverty, Building Solidarity, and Creating Jobs.' While the summit did not given much explicit attention to environmental issues, there are numerous references to the links between poverty, women's status, population growth and environmental degradation, all of which are intricately tied together in many less developed regions. The Copenhagen Declaration on Social Development made the following points in reference to the issue of linkages:

> *(6.) We are deeply convinced that economic development, social development and environmental protection are interdependent and mutually reinforcing components of sustainable development....*

> *(8.) We acknowledge that people are at the centre of our concerns for sustainable development and they are entitled to a healthy and productive life in harmony with the environment. (UN 1995B).*

Moreover, in the prescriptions for implementation and follow-up (Chapter V), the Copenhagen Declaration recommends:

> *The integration of goals, programmes and review mechanisms that have developed separately in response to specific problems.*

A Note on Terminology

The terms 'developed' and 'developing' assigned to countries have often caused difficulties and misunderstanding. Economic development can be given a precise definition: long term increases in real output per capita. But this is a variable, and the question of where the division lies between the more or less developed has often caused needless controversy. The term development or developed takes on a great deal more meaning, which is often less precise and open to controversy when applied to the political, social or cultural conditions of a society. The World Bank has provided one useful solution in ranking countries by their level of Gross National Product per capita, and then describing groups as low, middle or high income countries. The United Nations Population Division has another solution, which we shall adopt for the most part in this volume. Its documents speak of the More Developed and Less Developed Regions, and of the Least Developed Countries. The 1992 revision of *World Population Prospects* (UN, 1993) indicates that there are now 47 countries in this latter category, six of which were added since the 1990 revision, 'after the approval of the General Assembly in December 1991.' This phrase indicates as well as any that these definitions have important political implications. We use the More and Less Developed Regions classification for simplicity and because it has achieved some currency in population documents. We attach no moral, social, cultural or political significance to the terms, however.

All of these bases for action are reflected clearly and coherently in the basic policy document of IUCN, *Caring for the Earth*, which was published in 1991 and adopted by more than 80 countries (IUCN, 1991). At various points in the manual we shall draw on *Caring for the Earth*. It reflects the global bases for action, but also has a major advantage over them. The language of *Agenda 21*, for example, could not directly confront the issue of fossil fuels, and the need for cleaner, more sustainable sources of energy, nor could it say very much about population. The International Conference on Population and Development was able to say only very little about the population–environment linkages and what should be done. It was also less free to speak of the needs of women when these ran counter to specific religious positions. IUCN is somewhat less constrained by these understandable global political concerns. Its position can be based more firmly on scientific observations of the global condition. Thus *Caring for the Earth* can speak somewhat more frankly and openly about all of these needs and their interconnections.

Part I

Introduction

This introduction discusses use and users of the guide. It is designed to be used by national or regional (state, province etc.) level planners who wish to link population and environmental conditions more closely in planning for sustainable development.

The introduction also lays out the basic orientation, which is taken largely from IUCN's basic policy document, *Caring for the Earth* (IUCN, 1991). The aim of this document, and of IUCN, is to promote sustainable global and local communities. Sustainable development is defined as promoting the wellbeing of people and ecosystems. Although this volume gives emphasis to population conditions, it adheres to the basic position of *Caring for the Earth*, that both rapid population growth and high consumption are unsustainable.

Finally, the section provides an historical overview that brings more detail to the volume's basic orientation. The current conditions of global environmental change derive from the integral connection between population growth and technological and social change. The transition to fossil fuels, the rise of urban industrial society and rapid population growth are integrally intertwined, each causing and caused by the other. Gradual changes in the past have given way to exceptionally rapid and potentially destructive changes today. The growth of human populations and of human consumption are now unsustainable. Promoting sustainable development has become a widely shared responsibility of governments and peoples throughout the world. To promote sustainable development, we must build bridges between specialized disciplines and activities to treat our problems in a more holistic manner.

The Guide: Users, Use, and Basic Orientation

Users

This guide is designed primarily for two types of national level planners: those concerned with overall development planning, and those concerned more specifically with environmental conservation.

It attempts to show how population conditions can be taken into consideration in addressing problems of promoting economic development, environmental conservation or, more broadly, sustainable development.

A major aim of this volume is to show how population conditions can be examined and linked to planning for environmental conservation and sustainability. Much of the material on population necessarily emphasizes growth rates and the growth of numbers. This is, of course, primarily a condition of the Less Developed Regions, and arises in part from recent dramatic successes in controlling mortality. The current growth rates in these regions are not sustainable and they are closely related to pressures that reduce the wellbeing of both people and the environment.

Nonetheless, as we state at a number of places throughout this volume, the basic orientation taken here is derived from *Caring for the Earth* (IUCN, 1991): *both population growth and high consumption are unsustainable*. Both must be dramatically altered. *Caring for the Earth* provides many useful guidelines on ways in which production and consumption must change to promote sustainability.

This volume focuses more on population, however, which implies giving much attention to the condition of rapid population growth in the Less Developed Regions. There are three main reasons for adopting this emphasis:

1. It is where populations are growing rapidly that we can see some of the most visible linkages between population dynamics and the environment.

2. It is also where populations are growing rapidly that we can see most clearly the interlinked problems of poverty, inequality and the population–environment dynamic.

3. Perhaps of even greater importance is that the negative aspects of rapid population growth can be mitigated, far more easily than is often believed, by addressing the problem of growth directly. Today we have far greater capacities to intervene to reduce both mortality and fertility than we have ever had in the past. Moreover interventions for the control of mortality and fertility can greatly increase human welfare and the quality of life. As we shall note later, it is not as easy to intervene directly to control human migration, the third component of population dynamics. It can be predicted, and this is important, but it is far less easily controlled by direct intervention than either mortality or fertility.

Use

This is not a book of recipes to be followed mechanically. It is rather a set of ideas and options from which planners and implementers can choose activities appropriate to their specific conditions.

It can be used at national, state or provincial, and district levels, where planning covers a substantial geographic area with many different environmental conditions and many local communities. It can be used by government organizations, or by non-governmental organizations for project planning, or for monitoring the impact of government policies and programmes.

National planning bodies can use the guide to help search for and identify specific sectors or regions where population and environmental conditions pose specific problems, and where strategic interventions can be planned. National bodies can then be led to consider devolution of responsibility and authority for addressing any specific problem. Similarly, national level planning groups in specific sectors, such as

agriculture, forests, health, or urban systems can use the guide to identify specific activities where population and environmental dynamics appear to create problems, and then can plan strategic interventions to address those problems.

The guide can also be used at state, provincial or district levels, wherever administrators have responsibility for a substantial geographic or administrative area. For example, many national planning bodies or agencies now have parallel provincial and district level development planning units, whose local plans fit into and contribute to national plans and have special responsibility and authority for implementation at local levels. In such cases, the guide can be used at all levels, to assist in generating a comprehensive strategy for dealing with the population issues in development stimulation throughout the nation.

Basic Orientation

This work is based on the key policy document of IUCN, *Caring for the Earth* (IUCN, 1991) which provides both a visionary and a practical statement of what needs to be done to achieve sustainable global and local societies. It has been adopted as IUCN's basic policy, and has also been adopted by over 80 countries which are members of the Union. As noted in the preface, the basic message of *Caring for the Earth*, was reflected in both *Agenda 21* and in the World Programme of Action of the International Conference on Population and Development.

Caring for the Earth is subtitled *A Strategy for Sustainable Living*. Later we discuss the issue of strategies. Here we must address the issue of sustainability, since it is central to the vision of IUCN basic policy. The term is now commonly used in connection with both development and environmental conservation. *Caring for the Earth* provides the definitions

Definitions of Sustainability

Sustainable Growth is a contradiction in terms. Growth cannot continue indefinitely. The term is neither used nor to be inferred from the discussion in this handbook.

Sustainable Use refers only to renewable natural resources; it means using them at a rates within their capacity for regeneration.

Sustainable Development implies increasing human productivity and the quality of life while keeping within the carrying capacity of supporting ecosystems.

Source: IUCN, 1991, p. 10

Elsewhere (Carew-Reid 1994), sustainable development is defined more succinctly as:

PROMOTING THE WELLBEING OF BOTH PEOPLE AND ECOSYSTEMS.

Is National Planning Possible?

Failure of State Planning

It may appear incongruous to write of national planning for sustainability today, when central planning and state interventions are everywhere being displaced by privatization and the reliance on markets. To be sure, much central planning has turned out to be less than successful. It has nowhere lived up to the dreams that attended the early Soviet plans, or the central planning of many new nations that gained independence following World War II. Too often central planning has been associated with erratic oppression, corruption or bureaucratic obstruction. Planning and state intervention have been very much discredited, and today they seem to be in full retreat.

Global Forces

Moreover, the capacities of individual states to manage their own economies and societies are weakening under the pressure of global forces that they cannot contain. Global economic conditions intrude heavily on the economy of any state, and are for the most part beyond state control. International capital flows, controlled by a small number of large transnational institutions, have resulted in heavy debt burdens that now appear unsupportable in many poor countries. The power of external institutions to impose structural adjustment packages is all too evident. Transnational companies have far more power than do most states, to move everything from capital and people to toxic wastes. The breakup of large states, the emergence of ethnic conflicts and a massive global arms trade threaten all states with violence and refugee floods that can disrupt any planning. Under these conditions we can not expect a great deal of national level planning.

Possibilities of State Planning

But these conditions do not by any means negate the need for planning and the ability of states to create policies and programmes that can promote sustainable development. States still do have some capacities to control borders, to mobilize resources, and to direct those resources towards productive activities. They have capacities to establish policies that stimulate and enhance individual achievement, and collective welfare. Some of the most important policies and programmes concern the development of human capital, through promoting health and educational services; the protection of natural resources; planning for sustainable use; and the empowerment of local communities, through devolution of responsibility for resource management. It is in recognition of these possibilities that this volume is especially directed.

on page 4, which are those used in this guide.

We recognize that for many low income countries, economic growth is considered a necessity. The demand of poor people and poor countries to raise the standard of living and quality of life is a legitimate one and must be supported. As *Caring for the Earth* maintains, the high inequality of wealth in the world today is neither stable nor sustainable; gross disparities in consumption must be overcome (p. 44).

To the extent that economic growth is necessary to raise standards of living, especially amongst the poor, it must be supported. But growth cannot continue endlessly, and there is much evidence now that economic growth itself will not necessarily produce sustainable local and global societies. The authors of *Beyond the Limits* (Meadows, 1992) merely restate what the World Bank (1992) and UNDP (1993) have to say about the process of economic growth without increasing employment. Nor does economic growth necessarily by itself alleviate problems of poor housing, the homeless, poor health, or other conditions associated with a lower quality of life. Growth has thus proved in part a failure. To advance the quality of life of human beings and the world's ecosystems, to produce sustainable local and global societies, different models and different ethics must be found. The idea of sustainable development, for all its imprecision and weakness, provides a better vision than does economic growth.

Companion Volumes

IUCN produces a variety of planning guides for specific environmental conditions, including wetlands, parks and protected areas; for specific species; and for the broader vision of biodiversity. The full list of such guides appears in the appendix. There are two guides, however, that are of special interest to the problem of this guide, integrating population into strategies for sustainability. *Our People, Our Resources*, is a manual designed primarily for work at the local community level in rural environments. It offers options and resources to promote participative assessment and planning around the integrated management of natural resources and population dynamics. It also refers to 'Primary Environmental Care', or PEC, an approach to community management of resources promoted by IUCN and other organizations.

A second companion volume of particular interest is *Strategies for National Sustainable Development* (SNSD), by IUCN and IIED (Carew-Reid, 1994). It provides ideas, options, and a distillation of lessons learned from creating national level plans for promoting sustainable development, or National Strategies for Sustainable Development. (See Box on SNSD, p. 16). The ideas presented in this guide are meant to be integrated into planning and implementing those strategies for sustainable development.

Overview

The guide begins in Part I with a *review of the global historical context* of the population–environment relationship. It is necessary to recognize the distinctive context of the current global conditions of unsustainable growth and the current efforts to address the problem. These efforts are organized in specific ways in which national and international organizations attempt to intervene with human processes of production and consumption. A major and pervasive aim of these interventions today is to promote economic development, the long term rise of human productivity and human welfare. These global conditions and the organized interventions have emerged with an accelerating rate of change over the past two centuries, producing our modern global society.

Companion Volumes

Our People, Our Resources, A manual to assist rural communities in population appraisal and planning processes, Barton et al, 1996. *Strategies for National Sustainable Development*, A Handbook for Their Planning and Implementation, Carew-Reid, 1994.

It is increasingly clear that these past efforts at raising productivity and welfare have produced both a higher quality of life for much of the human species, and also major environmental stress, placing our planet at risk. This now presents an important new challenge for the world community. The challenge is to organize for planning and implementing strategies for sustainable development.

In Part II, the guide considers a series of *issues in organizing* at the national level for integrating population issues with strategies for sustainable development. To promote sustainable development, it will be necessary to bring together highly specialized information from a number of sources to a place where it can be effectively used. Population, agriculture, forestry, finance, industry, health, transportation, education and pollution are only some of the important sources of knowledge and information that have become highly specialized, both in scientific disciplines and in large scale government and private agencies. The specialization has been important and has increased powers of observation and of planned intervention. But the specialization has also built barriers, making communication difficult. At the same time, all of the real world problems we face tend to spill over the boundaries of specialized activities. They need to be brought together to focus on human problems, rather than upon their specializations alone. This is very much an organizational problem. It requires drawing together the powerful *specializations* that are an integral part of the way we allocate resources and make collective decisions. The very fact of specialization implies a resistance to integration, and special tactics are required to overcome this resistance. In this part, we provide some suggestions for organizing to build bridges across the boundaries of specialized scientific disciplines, and specialized agencies. Each

section in Part II includes a set of questions concerning population and environment dynamics, and the organization of linkages, which will refer the reader to tools discussed in Parts III and IV, where we provide a series of suggestions for assessing population–environment dynamics.

Part III focuses specifically on the population–environment nexus. It *reviews the major frameworks and models* people have used to think about the population–environment link, and pays special attention to a powerful modelling process, the Population Environment Development Model, produced by the International Institute for Applied Systems Analysis (IIASA). This section also reviews some of the major types of linkages, suggesting different tools, from which governments can choose to work on what we find are basically local, or location-specific problems.

Part IV of the of the guide is something of a primer on population, with some discussion of environmental measures as well. It lays out the *tools and perspectives of population analysis* that will be needed for strategies of national sustainable development. It begins with a discussion of the Demographic Transition, one of the most profound changes the human species has experienced. The Demographic Transition is often misunderstood, however, and taken to indicate that little can be done to affect patterns of human reproduction. This misunderstanding is much in need of correction, since today we find that human reproduction can be changed far more easily than is often believed. Moreover, promoting that change is important for increasing human quality of life, and especially for promoting the wellbeing of women. This section also reviews basic demographic tools, as well as the course of population dynamics especially in the low income countries of the world.

The Global, Historical Context of Population–Environment Dynamics

Population, Environment, Technology and Social Organization

Figure 1 shows the trend of the human population over the past 1000 years, in what is probably one of today's most common pictures of the global scene: centuries of slow growth, with an explosion of numbers only in the past half century. The growth is closely associated with increasing environmental stress. The major trends in global environmental change include the following:

- atmospheric changes that threaten to increase the global temperature through the accumulation of heat-trapping carbon dioxide and other 'greenhouse' gases, and increase ultraviolet radiation reaching the earth through stratospheric ozone depletion;
- pollution of the air, water and land through the release of toxic chemicals and waste materials;
- ecosystem degradation such as deforestation, soil erosion and salinization, eutrophication of water bodies, loss of wetlands and erosion of coastlines;
- loss of biodiversity through habitat destruction; and
- the increased potential for global life destruction through nuclear and chemical weapons of mass destruction.

There have been radical global changes in the earth's biological functions in the past and radical changes may be expected in the future. But never before has one species been able to have the kind of global impact that the human species has today.

Estimates are that the human species now appropriates 40 percent of the total primary production of the earth (Vitusek 1986). This has led some ecologists to speak of the human species as a deadly epidemic on the biosphere. Others speak of a population bomb (Ehrlich 1972) that threatens human and possibly all life on the globe.

Figure 1 is deceptive, however. It suggests that it is population growth that is driving global environmental degradation. Figure 2 provides two pieces of additional information crucial to understanding the current dynamics of population and the environment.

The shift to fossil fuels in the past two

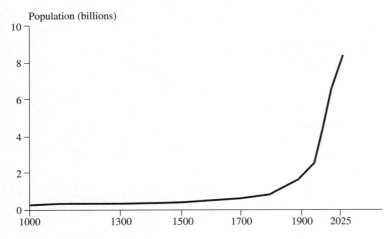

Figure 1 *World population, 1000–2025*

centuries has given us a new type of society and technology — urban industrial society. Fossil fuels increased human productivity and transportation capacities, thus promoting population growth, urbanization and environmental change (Boserup 1965 and 1981, Livi-Bacci 1989). The fossil fuel technology itself, however, did not spring out of nothing. It was also the product of prior population dynamics and social changes. Thus the shift to fossil fuels is as much a result as a cause of the new urban industrial society. The major lesson we take from this is the following:

Population growth, urban industrial society, economic development, environmental degradation and loss of biodiversity form a seamless web. Each is both a cause and an effect of the other. None can be effectively addressed in isolation from the others.

There are also two important corollaries of this observation:

There is no single, simple and direct relationship between population and the environment.

All population–environment relationships are mediated by some form of human technology and human social organization.

Two Broad Historical Patterns

Two broad patterns can be seen in the population–environment relationships over the past two or three centuries. Earlier changes were more gradual and more geographically limited. Current changes of our modern urban industrial world have accelerated rapidly and have a more global impact. A major problem today is dealing with the speed of the change, and the globalization of the impact.

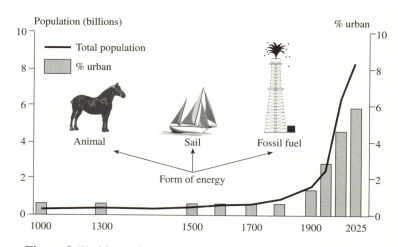

Figure 2 *World population, energy and urbanization, 1000–2025*

Gradual Changes of the Past, and the Birth of a Global Community

Until about 1950, the expansion of food production came largely from the physical expansion of land under cultivation. This often required much environmental change, for example in the construction of irrigation systems, and the cleaning of forests. But yields per area remained more stable, often showing consistency over centuries. This implied relatively gradual increases in food supplies, often linked to the development of new lines of transportation.

Traditional Rice Yields

On the Wen river in China, the basic irrigation system that today waters the rich rice fields was laid down 2000 years ago. Until recently, the long term output of rice in the region was about one to two metric tons per hectare.

Millions of hectares of new rice fields were carved from the tropical forests in Southeast Asia from roughly 1850 to 1950. Typical rice yields throughout this period were about one to two metric tons per hectare.

Similarly, past increases in population tended to follow gradual increases in the standard of living, and the expansion of basic public health infrastructure, with relatively little impact from new medical technology. The decline of infant mortality, for example, from high traditional to low modern levels, usually required one to two centuries.

The expansion of ocean transportation from as early as the 15th and 16th centuries tied the world together in a single human-dominated ecosystem. New crops came out of the Americas to increase the carrying capacity of the land in Asia, Africa and Europe. In the other direction, diseases new to the Americas came from Europe and along with new forms of political oppression and slavery caused major demographic collapses among indigenous populations in Central and South America. Though changes could still remain limited in their geographic scope, a new community was emerging, which would soon connect all parts of the globe.

The gradual expansion of industrial activity, starting in Europe in the late 18th century, brought increasingly radical changes in the global population–environment relationship. The demand for natural resources and for food brought widespread environmental change. Forests were cut for timber and turned into agricultural land, mining drew minerals from the earth, often leaving behind wastelands. At the end of the 19th century all of these trends accelerated with the invention of the internal combustion engine and the extraction of oil to fuel the rapidly accelerating, fossil fuel driven, urban industrial society.

A new global community was slowly arising on the basis of the urban–industrial society. Western imperial expansion forcefully drew traditional lands and peoples into a new market-oriented world society, organized more and more by modern forms of bureaucratic organizations. Governments and markets increasingly drew isolated local communities into a global community. The population–environment dynamics of the past brought the gradual development of a worldwide urban industrial society, whose environmental degradation now appears to be unsustainable.

Acceleration of Changes in a More Integrated Global Community

Since 1945–50, all the gradual changes of the past have accelerated. Changes in both technology and human social organization have produced a radical break with the slower movements of the past. New forms of energy, new machines, new organizations of science, industry and commerce, new national political movements, and the development of new global organizations (intergovernmental, non-governmental and business) have brought a population–environment relationship that is unsustainable, presenting a major global challenge.

Our modern global society is *unsustainable*. We have created a self-reinforcing combination of rapid population growth and high resource consumption, both of which are unequally distributed throughout the world. It is both the extreme levels and the inequality of population growth and resource consumption that make the current global situation unstable and unsustainable.

Evidence of the unsustainability of our levels of consumption and population growth are not difficult to see. Current population growth rates continuing for another 300 years would give us a world population of not 5 but 1500 billion people; continuing for a mere 2000 years would produce enough human beings to cover all known matter in the universe. No rate of population growth is sustainable. The rate of population growth in the more developed regions is currently only about 0.4 percent per year (UN 1994), but since one inhabitant of the more developed world consumes much more than one in the less developed world, even the slow level of population growth makes a far greater impact on the global environment. Per capita energy consumption in the US is twice that in Japan or Europe, but more than 26 times as high as Africa. Energy consumption to support a projected world population of ten billion at a conservative figure of 7.5 Kw per capita would require five times the total energy consumption of the current world's population (Daly, 1994). Today residents of the wealthy countries require four to six hectares of land per person to support their high levels of consumption. If the entire world's population were to consume at this level, it would require twice the actual land area of the entire globe, or more than three times the current arable land area (Daly 1994). Such illustrations can go on almost endlessly. All lead to the inevitable conclusion that current rates of population growth and of consumption are unsustainable.

The new forms of human social organization arising in the global community also signal a radical change from the past. The western overseas imperial system has given way to over 150 independent national states. New forms of intergovernmental organizations, non-governmental organizations, and international business organizations have grown rapidly in the past half century. Both national and international organizations increasingly seek to intervene to promote economic development and social welfare, to protect the environment, and most recently to promote sustainable development. Large scale national development planning came to dominate the agenda of national and international organizations. Planning has weakened most recently (see Box,

---

Rapid Modern Change

- The infant mortality decline from high, traditional to low, modern levels required two centuries in the west. Chinese infant mortality declined from 195 per thousand in 1950 to 32 per thousand in 1990.
- After 1965, rice yields in many regions rose from one to four tons per hectare in less than 20 years.
- Commercial energy consumption grew by about 2 percent per year from 1850 to 1950; it has grown by 4.5 percent per year since 1950.
- World population growth rates remained under 1 percent per year 1850–1950; they rose to 2.2 percent by 1965.

---

p. 5) but still remains a common activity.

In the new national planning, however, the different areas of population, environment and development are typically highly specialized and often divorced from one another. Forest protection is often planned without reference to population growth in the surrounding area. Basic health services are promoted without reference to education or income generating activities. Agricultural development is promoted without reference to population growth, long term and competing water requirements, or pollution caused by fertilizers and toxic chemicals. Urban planning goes on without reference to competing water requirements, agricultural development, or population growth. Family planning works to promote fertility decline without addressing issues of health and social welfare or the distinctive status of women. The list could go on and on. The results are typically weakness at the project level, and sometimes real disaster as well.

Three lessons can be taken from this broad historical review.

1. The growth of human populations and resource consumption has become unsustainable.

2. Governments and both national and international organizations will attempt to intervene in planned efforts to promote sustainable development.

3. Success in promoting sustainable development will require broad strategies that tie together a variety of issues dealing with economic development, population growth and environmental conservation.

Part II

Population–Environment Linkages in Strategies for National Sustainable Development

Making effective linkages between population and environmental issues in promoting sustainable development will be in large part an organizational process. It requires building bridges between specialized activities: between scientific specializations; between the specialized units of government agencies or non-governmental organizations; and, above all, between individual specialists.

This section reviews some of the organizational problems that must be addressed. These include the basic idea of sustainable development and its current imperatives, the rationale for making the linkages, and the advantages and disadvantages of making linkages. The concept of political–administrative strength is introduced to address the question of why some communities (or governments) do better than others in transforming wealth into welfare. It also addresses important organizational issues that come under the rubric of what has been called *integration*, which is a process that has engaged development organization planners for years and has involved a great deal of confusion. It is also, however, an area where there are good experiences on which to build. The chapter concludes with a specific organizational strategy for creating Population Environment Networks as a strategy for bridge-building between specializations.

At each of the sections in this part, we raise a series of questions, suggesting where the linkages are, how they can be made and what tools can be used to assist in making the linkages. We also suggest specific measures that can indicate strengths and weaknesses of the organizational tools planners typically work with. Often these questions will refer the reader to the specific tools that are included in parts III and IV of the guide.

Strategies for Sustainable Development

Strategies

Strategies for sustainable development are relatively recent phenomena, but they rest on a complex set of developments that have evolved over the past four decades. The first is national economic development planning, which has been a major activity of most of the low income countries, especially those that emerged as new independent states after World War II. Development plans have been promoted by the World Bank and various international aid agencies. In 1960, the United Nations Development Programme (UNDP), launched the decade of development and greatly promoted the idea of national development plans. Using national income accounting and based on economic theories about capital mobilization, these have become elaborate and sophisticated (though not always effective) tools for planning at both national and international levels.

Shortly after the first decade of development was launched, disturbing signs of environmental stress began to be seen. Rachael Carson's *Silent Spring* (1964), was one of the first popular statements to sound the alarm of increasing pollution. In 1972 another significant book, *The Limits to Growth* (Meadows, Meadows and Randers) sounded another alarm. A series of computer simulations from a world systems model of development suggested that current patterns of worldwide growth in consumption, industrial production, and population were unsustainable. The 1972 Stockholm Conference signalled both worldwide conflict and consensus over the impact of the then standard models of economic development. The decade of the 1970s saw a new awareness of environmental stress caused by economic development.

This decade was also one of rising awareness of a population problem. The first International Conference on Population was held in Bucharest in 1974. Like the Stockholm Conference, it was attended by serious disputes, especially over the global inequality in wealth, but it ended with a World Plan of Action that sought to reduce the rate of population growth by focusing both on fertility limitation and on related development aims.

Programming and international assistance for national economic development remain major activities in the world today. They are increasingly criticized, however, for their heavy emphasis on capital accumulation for industrialization, and for their failure to place sufficient emphasis on human welfare and poverty eradication. They are also criticized for their failure to assess natural resources properly (Repetto, 1989), and to count the environmental costs of natural resource use and industrialization. Moreover, they rarely give proper attention to the problems of rapid population growth, and its links to poverty, inequality and women's status.

There have also been positive developments in the area of environmental conservation. National conservation plans,

Strategies for National Sustainable Development (SNSD)

1. Strategies for sustainable national development seek to improve and maintain the wellbeing of people and ecosystems: A strategy is a process by which groups design and take a set of actions:
    *   to strengthen and change values, knowledge, technologies and institutions;
    *   to achieve specific objectives; and
    *   to improve and maintain the wellbeing of people and the ecosystem.
2. The overall goal is sustainable development.
3. The choice of strategy objective should be strategic. Location-specific conditions will determine which specific goals should be chosen. These may be broad national goals, or where capacities are more limited, they should be narrower goals that can in fact be achieved.
4. The strategy process is cyclical and adaptive. A strategy process is a continuous activity of planning, acting, reviewing, revising, planning and acting.
5. The strategy should be as participative as possible. Wide participation strengthens both the planning and the implementing of a strategy.
6. Communication is the lifeblood of a strategy. Communication permits knowledge to be shared, values to be changed and strengthened, and adaptations to be made to achieve the goals of a strategy.
7. Strategies are processes of planning and action. Strategies imply both planning and implementation. They are processes of developing a long term vision, and taking steps to realize that vision.
8. The strategy should be built into the decision making processes of the society as a whole. Strategies should be built into national development plans and into local processes of planning and implementation.
9. Building capacities to undertake a strategy should be done at the earliest possible stage. Creating a strategy requires capacities, which can be learned at the start.
10. External agencies should be on tap not on top. External financial and technical assistance is often needed to create and activate strategies, but it must not dictate or dominate the process.

Source: Carew-Reid 1994

stimulated in part by the 1972 Stockholm Conference, have gained currency. A milepost in this development was the publication of the IUCN's *World Conservation Strategy* in 1980. This called upon governments to develop national and subnational conservation strategies. *Our Common Future*, the 1987 report of the Bruntland Commission, called for the same thing, and this was further supported by a joint IUCN/WWF/UNEP 1991 publication, *Caring for the Earth*. More recently *Agenda 21*, the Report of the United Nations Conference on Environment and Development in 1992, registered wide agreement on the need for sustainable development.

Strategies for sustainable development thus reflect the growing recognition that we need a new kind of development which promotes the wellbeing of both people and the environment. The strategies also reflect the recognition that achieving such wellbeing will require linking population, development, and the environment in integrated strategies for action.

Underlying this integration strategy for sustainable development, however, are five fundamental propositions that must be made explicit. They derive from *Caring for the Earth*, but we believe they are also coming to be widely accepted in the larger agenda of development.

The Meaning of Development

Drawing from the orientation and arguments of *Caring for the Earth*, and from the *Strategies for National Sustainable Development* (Carew-Reid, 1994), we can identify five fundamental propositions that underlie this attempt to integrate population into strategies for sustainable development. Since all five refer to 'development' it will be useful to begin with a brief statement of what we imply by development.

Development, here and in *Caring for the Earth*, implies economic development. For this there is a very precise definition – the increase in human productivity. Economists have given us a useful operational definition of this as long term increases in real output per capita. National income accounting has given us a useful tool for assessing output. We have seen that what we now call economic development has been historically associated with an increase in the quality of human life. Infant and total mortality rates as well as a whole range of infectious disease rates have declined, and education and literacy have increased, along with many of the kinds of resource needed to give human beings the opportunity to develop and exercise their distinctive talents. This positive character of economic development is also found in the differences today between rich and poor countries. All quality-of-life measures are closely related to the level of economic development. In this respect economic development, or the increase in human productivity has been and is a good thing. It is no wonder that most countries today are strongly committed to achieving economic development.

This perspective on economic development, and especially national income accounting, has, however, increasingly come under attack for the real distortions it causes (Repetto 1989, Daly 1992, Kallenberg 1993, National Research Council 1994). When we count only goods and services that gain monetary value by flowing through a market, we omit much of human productivity, especially that of women. We also add to wealth much activity that degrades rather than enhances life; cleaning up a massive oil spill, for example, adds to the national wealth. When we value as capital only man-made capital, we count natural resources only as income when they are depleted. That total deforestation or cleaning up a massive oil spill adds to a country's GNP is not the way to think about or plan for sustainable development.

Economic development may mean economic growth, but growth is not sustainable. Moreover, as many have noted (Meadows, Meadows and Randers 1992), economic growth has not solved many of our problems. Growth has not solved problems of homelessness, poverty, crime, drug abuse, and violence, nor has it even solved problems of unemployment as today we face a new problem of economic growth without jobs (UNDP 1993). By development then, we refer to an increase in human productivity that is sustainable and that enhances, rather than degrades, the quality of both human life and ecosystems. We believe that this type of development – sustainable development – is possible, and it is this type of development we refer to here.

To link population with this type of sustainable development, requires the elaboration of five fundamental propositions which deal with development and its benefits. More importantly, they deal with the responsibility of national governments to help their citizens achieve the benefits of development. Thus they underlie our perspective on how population is to be linked to strategies for national sustainable development.

1. Development is an imperative

All countries today espouse the major aim of promoting their own development. Development means increasing human productivity, and only by increasing human productivity can nations provide for their citizens a decent standard of living and an opportunity to develop their talents. It is a major responsibility of national governments to promote their own national development.

A large portion of the world's population (estimated at 1.2 of the 5.7 billion) today suffers from the disease of poverty, consuming far less than their share of the world's resources. A minority of the world's population enjoys high standards of living and consumes more than their arithmetic share of the world's resources. As we have said before, this situation is unstable and unsustainable. It is imperative that the poor nations succeed in achieving substantial development in order to produce a sustainable global society. The world community, especially the wealthy nations, have a major responsibility to assist poor nations in achieving substantial development.

2. Development is a means to improve the quality of life of people

Development is a means, not an end. The end is a quality of life that sustains human life in comfort and dignity, enabling people to grow and to develop the talents they have as distinct human beings. National development and the wealth it produces are necessary for this, but development is not an end in itself.

Development and wealth do not always translate into human welfare. At every level of economic development, some countries or communities enjoy much higher levels of welfare than others. National policies and capacities for effective planning and implementation are some of the major conditions that determine the amount of welfare a population derives from its national wealth. It is the responsibility of national governments to translate levels of wealth into the highest possible levels of human welfare.

3. Development is for future as well as present generations, therefore it must be sustainable

Development plans and projects often fail because they do not meet human needs and yet destroy or degrade the natural resource base. Both present and future generations suffer. Development and conservation are not in opposition to one another. They are inseparable parts of sustainable development. Development without conservation robs future generations of welfare, but conservation without development equally robs current generations of welfare. National development plans and projects must have a sound base in environmental conservation if they are to produce sustainable development. National governments are responsible for promoting sustainable development.

Where national governments lack the capacity to promote sustainable development, it is the responsibility of the world community, and especially the wealthy nations, to assist in building that capacity.

4. Planning and implementing strategies for sustainable development requires linking together a wide range of specialized skills

National governments direct scores of highly specialized organizational units: agencies for agriculture, irrigation, forests, road construction, industry, health, and education, not to mention police, courts and national defence. The list is very long. Specialization enhances skills and increases the capacity of an individual or an organization to do something and to do it well. But specialized units derive strength from doing only limited things, while promoting sustainable development requires doing many things in concert with one another. Promoting sustainable development implies that agriculture must be able to produce sufficient food for today and for the future. This in turn implies that agricultural development without environmental considerations may well fail in the long run. Development often means industrialization, but without attention to pollution control, industrialization can be dangerous for future populations. Both development and welfare plans must take into account the numbers of people who will supply the labour and demand the services in the future. In many cases population growth undermines a nation's capacity to invest for development or for welfare. Planning effectively for sustainable development requires linking together the specialized development, environment and population agencies of government.

5. Effective planning and implementation in population and sustainable development requires broad popular participation

Participation has become a key word in all development activities. National governments, especially in newly independent countries, rushed to take control of development processes in the first two decades after World War II in the belief that local initiative was weak, social and physical infrastructure was underdeveloped, and that there would be no movement without government intervention and initiative. We are now quickly learning that governments cannot act alone. Any successful development plan needs broad participation if it is to succeed.

There are three major benefits to be derived from broad-based popular participation.

a) More knowledge and information about local needs and resources will be put into the planning and implementing process;
b) adaptation to local conditions will be better; and
c) more effort will be mobilized by local people and groups committed to the achievement of the plans.

Additional benefits can also be expected by increasing the capacity of local groups to promote sustainable development; often, too, this leads to increased trust in and support for the government. Promoting and permitting popular participation is not easy, often because governments have, in the past, tended to discourage such participation. It is the responsibility of the central government to promote and permit popular participation in activities to link population with strategies for sustainable development.

In the five fundamental propositions, we have spoken much of the responsibilities of both national and international governments. We believe that governments do have such responsibilities. Moreover, it appears that when governments do accept and discharge those responsibilities effectively, they gain popular support, and do not need force to stay in power.

Much popular resistance to government today comes from resentment of the abuse of power, and of the failure of governments to accept and discharge responsibilities effectively.

Thus, along with the responsibilities, there are also benefits for governments. When governments deal with a broad range of human problems, encourage popular participation, are sensitive to the demands of the governed, they gain popularity, loyalty and respect. When they are willing to deal with difficult and intractable problems, providing effective leadership in addressing controversies and helping to find ways to manage and resolve controversies rather than to quash them, they become stronger, more popular and more effective. One of the major controversies governments often face is that over population policies. Before moving further into population and environment policies, we should consider some of those controversies, and also some of the misperceptions of the population issue.

Thinking About Linkages

Necessities and Problems of Linkages

It is important to recognize that there are basic necessities that lead us to attempt linking population and the environment in promoting strategies for sustainability. But it is also important to recognize that there are difficulties in this linkage as well.

Why Try to Link Population and the Environment?

1. Because they are integrally related in life.

The 5.7 billion people living on the Earth today represent numbers and growth rates that are unprecedented in the history of the earth. Growth rates are certainly unsustainable; they cannot continue.

The numbers and growth rates represent a great success and an equally great challenge. Economic development and the transition to urban–industrial society based on fossil fuels – is responsible for the success of the species, reflected in its growth and dominance.

The success has brought, however, massive pollution of the earth, air and water, habitat destruction, and loss of species; in fact a level of environmental degradation that threatens human, and possibly all, life on Earth.

Numbers, success and threat are integrally linked to one another. They are part of the seamless web of the modern society we have created.

2. Because their problems cannot be effectively addressed in isolation from one another

The natural environment cannot be protected with fences and armed guards. It will only be protected by people who have a stake in its protection.

The health and welfare of people cannot be promoted in an environment that is degraded. They require a clean, rich and biologically diverse environment.

Human poverty cannot be eradicated by economic growth that destroys the natural environment and enriches only a few.

Human population growth must eventually end, but it cannot be curbed by coercive government population policies, nor should it end through a rise of mortality.

3. Because they can be effectively linked to promote a more sustainable society

Development programmes can succeed when they promote popular health and welfare, especially of women and the poor. This, combined with family planning programmes, can lead to population stabilization. Promoting human health and welfare can lead to attitudes and behaviours that are environmentally protective.

Environmental protection can be promoted when it enhances peoples' long term survival, and especially when they see direct benefits for themselves and for their children.

See Part IV for the full range of population measures, and a select range of environmental measures.

Questions

Here we should begin with the most general questions of the country's recent history of population dynamics, development, and environmental change. Using the past 20 years and looking ahead to the next 20, planners can ask such questions as the following. They will provide a broad and general view of the wellbeing of both people and ecosystems.

Population Dynamics

What is the current size of the population? How many people have been added in the past 20 years? How many people will be added in the next 20 years?

What is the crude death rate, the infant mortality rate, and the maternal mortality rate? What are projected changes in mortality rates?

What is the total fertility rate, and how much has it changed? What is the contraceptive prevalence rate? What are the projected changes in fertility and contraceptive use?

What is the net migration rate for the country as a whole? What is the rate of urban growth, compared with national population growth? What are the projections for urban growth?

Development

What is the level of GNP per capita? How much has it changed? What are projected changes?

Which sectors are the most dynamic, growing the most rapidly? To what extent is this due to internal investment policies, and to what extent due to external demand for products? What projections are made for the next 20 years?

Environmental dynamics

What is the extent of agricultural land degradation? How has this changed? What projections are made for land degradation?

What is the rate of deforestation, and how has this changed? What projections are made for deforestation, and for afforestation?

What is the country's annual freshwater availability? What is the annual withdrawal of water? What is known about water quality and pollution from human and chemical wastes? What are the projections of water requirements, and water quality?

What monitoring is done of air quality? What proportions of the population are subjected to substandard air quality? What are the projections for air quality?

Linkages

For all of these questions, we wish to know the connections: to what extent is population growth increasing demand for water, reducing water quality, increasing demand for wood and reducing forest cover, increasing demand for food and increasing land degradation? Or to what extent is environmental degradation reducing the health of a population, or forcing the population to move? For these linkages and many more one can use a series of simple frameworks, or a more sophisticated quantitative simulation model.

Problems and difficulties

1. Organizational communication and cooperation are difficult

Activities like population and environmental protection are dealt with by specialized organizations – units of government, non-governmental organizations, and scientific disciplines.

See Part III for examples of simple frameworks that can be used to identify and assess linkages, and a recommended simulation model that can be effective in planning.

These develop boundaries, internal identities, languages and technical tools that make it difficult for them to communicate with one another.

Often such units are locked into competition for scarce resources – power, influence, personnel, supplies and money. Members often work on the assumption that resources for one unit mean less for others. This makes it difficult for organizational units to cooperate with one another. Attempting to bring together these specialized units can lead to inter-agency conflict, and can reduce the capacities of specialized agencies to do what they were originally designed to do.

2. Population is often controversial

Promoting development is widely accepted because it means increasing wealth and possibly welfare. Protecting the environment is widely accepted by people attached to 'nature' and attractive forms of wildlife, though there are often conflicts between development and environmental protection. Population, on the other hand, has become the arena for controversy over issues of religion, ethnicity, morality, sexuality, and gender roles. Often these controversies are deep and seemingly irreconcilable, like the current debate over abortion. Often population issues involve ethnic identities, where conflicts can be deadly and longlasting. Many environmental organizations resist addressing population issues because they do not wish to be drawn into these intense conflicts.

These dangers and difficulties can be overcome, but they require good leadership and participation by a wide range of actors at all levels of government.

Common Misperceptions

While it is important to recognize that controversies exist around population issues, it is more important to note the degree of worldwide consensus that has been achieved since the world community actively began discussing these issues.[1] The International Conference on Population and Development in Cairo reflected both the past controversies and the emerging consensus (Ashford, 1995). Although there was unprecedented controversy involving religious organizations, there was also an unprecedented consensus, with the Vatican joining for the first time. The new consensus is built on the acceptance of a broad-based approach to population problems. The approach now accepted must include poverty alleviation and economic development, greater provision of social services – especially health and education for women – as well as attention to environmental protection. The World Programme of Action resulting from the Cairo conference was adopted by 180 countries.

Some of the past controversy over population issues was built in part on misperceptions of the problem. The following appear to be some of the more important misperceptions.

1. Population growth is THE major problem for sustainable living

This formulation tends to pit South against North in unnecessary conflict. Population growth is certainly a problem. Both numbers and rates of growth of the human population are unprecedented and unsustainable. But modern population growth cannot be separated from the total world condition: the rise of fossil fuel based urban industrial society with high inequality, grinding poverty for hundreds of millions and affluence and waste for a minority. Population growth as a problem for

1 That discussion may be said to have begun two centuries ago when Thomas Malthus wrote his first essay on population in 1798. There has been much discussion of population in scientific circles and in some political ones as well. But we can also take 1965–66 as a watershed period, since this is when United Nations (UN) bodies declared population to be a critical issue in world security and economic development. In 1965 the UN Economic and Social Commission for Asia and the Pacific, ESCAP (then called ECAFE, or the Economic Commission for Asia and the Far East), declared that population should be included in programmes of economic development support. The following year the UN General Assembly made roughly the same declaration.

sustainability cannot be separated from conditions of consumption, waste, inequality and poverty.

2. Population growth will slow only when urbanization and economic development have been achieved

This misperception rests on a misunderstanding of the differences between past and present changes in both death and birth rates. These differences are related to the Demographic Transition, discussed below in Part IV. What the differences between past and present tell us is that modern population dynamics, especially death and birth rates, can change much faster today than in the past. Both are strongly affected by what governments do to provide effective social service delivery systems, especially to the rural areas. Many countries today (e.g. China, Sri Lanka and Thailand) have achieved low levels of mortality and fertility even while money incomes are low and a majority of people still live in rural areas. Building effective delivery systems for education, health and family planning can help reduce both death and birth rates.

3. Population growth will slow when family planning services are available

Family planning programmes can certainly assist in reducing fertility, as we will show below. But providing contraceptives alone, without increasing services for health and education, or without increasing opportunities for families to improve their welfare and that of their children, will often lead to disappointment and controversies over the sources of continued high fertility.

4. Slowing population growth by reducing Less Developed Regions' fertility will solve

the problems of sustainability

Reducing fertility will certainly help in many ways to improve both individual and ecosystem health, but there are two important caveats.

1) Without addressing problems of poverty, inequality, and over-consumption, reducing human fertility will still leave us with an unsustainable society.
2) Population momentum (see page 114) implies different time horizons for fertility reduction to have an impact on other conditions. Even a rapid reduction in fertility will not slow actual population growth in numbers for ten to thirty years into the future. It will not reduce demands on land and labour for up to twenty years. It will not reduce demand for educational increases for up to ten years in the future. Reducing fertility does, however, have an almost immediate impact on improving maternal and child health.

Questions

At this point planners must address some of the more sensitive or delicate issues in population–environment dynamics. What is the national population policy? What are the reasons given for those policies? What groups support or oppose policies or suggested policy options? These questions should be asked about the full range of population issues: mortality, fertility, internal migration and urbanization, and external migration.

Note that the questions of policy have been asked in a systematic and periodical manner by the UN Population Division since this was mandated by the World Plan of Action adopted at the 1974 International Conference on Population, held in Bucharest. Currently standard questionnaires on national population policy are distributed to governments every two years, and are brought together in a *Global Population Policy Data Base* (UN 1995C). The UN publication can be used to identify the country's specific population policy, but additional questions should be asked about support and opposition to specific elements of the policy. For example a simple chart, such as the following, can be used to identify positions. These will be delicate questions and must be handled sensitively. The point here is to look for common ground to produce population policies that promote human wellbeing. We can assume that most if not all groups wish to promote human wellbeing, though they may

ISSUE	SUPPORT		OPPOSE	
	Group(s)	Reason(s)	Group(s)	Reason(s)
Mortality control				
Fertility control				
In-migration control				
Out-migration control				
Regional migration				
Urbanization				

differ in how to achieve this end. Making positions on the delicate issue of population policy is a first step in finding common ground.

Organizing Linkages

Political–Administrative Systems

All states today construct large scale administrative systems: to collect taxes, provide physical protection, promote communication, construct and maintain a modern physical infrastructure, and to provide goods and services. This system typically becomes highly differentiated and specialized, with different agencies or units assigned different kinds of tasks. These are, in effect, organizational tools that the human species has invented to do its work. They are, however, imperfect tools, and they are tools with different specific capacities.

It is quite obvious that these political administrative systems differ greatly in their capacity to do anything and everything. Some communities enjoy a high level of public health and educational services, are secure in their persons and property, can move about easily and be in communication with the entire world and enjoy a standard of living that permits them to develop their own talents. At the other extreme, some communities face constant violence, lack even basic services such as adequate clean water, and can expect little in the way of health or educational services. Most governments have aims and aspirations, and often even specific plans, to provide physical infrastructure and social services to their populations. There is a great deal of difference, however, in the capacities of their political administrative systems to actually deliver infrastructure and services.

Many of these differences in capacity are related to the level of economic development and national wealth. Wealthy countries can generally provide more security, better infrastructure and more social services to their populations than can poor countries.

Economic development and national wealth is only part of the reason, however, and perhaps not even the most important part. Sri Lanka and Costa Rica have more effective and service oriented political administrative systems than many countries that are much wealthier. At all levels of economic development (indicated by GNP per capita) we can find countries that differ in social services and in the quality of life provided for the population. Sri Lanka has exceptionally high levels of health and education, despite its relatively low GNP per capita, and its low level of urbanization. Similarly, Costa Rica has for some time defined itself as a welfare state, not a garrison state. Consequently its political–administrative systems give high priority to social services rather than to military matters.

What makes for high political–administrative capacity? How can a country improve the capacity of its system? And what impact does the capacity of the system have on the quality of life, the protection of the environment, or on population dynamics? These are important questions to be asked of planners developing a strategy for sustainability.

While strong and consistent answers are not readily available, there is one useful observation to be made about developing the capacity of a political administrative system. One of the strongest approaches to improving the capacity of a political administrative system is to promote decentralization and local participation.

One of the major tasks of a political administrative system also gives us a measure of performance: delivery of services. Many less developed regions find it difficult or impossible to provide even the simplest of services. Planned educational systems are often without classrooms, teachers or supplies in the rural areas. Primary health services often do not reach the rural areas. Getting doctors, nurses, medical supplies or educational materials to rural areas represents administrative, financial, personnel and logistical problems that many poor countries find great difficulty in solving. Thus the first task of a political–administrative system is to have its specialized agencies do what they are supposed to do. This often means not integration, but more and better specialization.

In a review of population and health delivery systems, Finkle and Ness (1985) observed great differences in capacities for service delivery and came up with a simple proscription: pay attention. They observed that family planning programmes can be found in a wide variety of organizational conditions. Some are freestanding, specialized programmes. Many are set administratively in the ministry of health, some are set in other government agencies. What makes a difference in performance is not where they are located, but whether an individual officer and a specialized organizational unit is clearly responsible for the delivery of services, is given the authority and resources to discharge its responsibility, and is held accountable for performance. Family planing is the responsibility of the health ministry in most Indian states. The result is not encouraging. It is located in the Ministry of Public health in Thailand and in the Ministry of Health and Social Affairs in South Korea. In both cases this works well because there is a special unit

The Cycle of Local Administrative Incapacity

The Cycle

For decades actors in and observers of development plans and projects have noted a problem in relations between central and local governments which persistently weakens local capacities to implement development activities. Central governments wish to promote projects and to see results as rapidly as possible. They tend to fear that local authorities lack the experience and skill needed to implement projects, thus they keep both control and financial resources in the centre. This directly slows down project implementation, due to increased transportation and communication demands. It also often produces inefficiency or failure, as the central project is not adapted to local conditions. Most important, central control denies local authorities the experience by which they gain implementing skills, thus requiring more centralization, in a vicious spiral that keeps the local units weak.

Breaking the Cycle

Decentralization, devolution of authority and allocation of funds can promote project implementation by increasing the participation of both local government agencies and local people. As with People's Participation, this can produce the following benefits:

- fuller use of local knowledge and skills;
- more efficient and effective project implementation;
- better adaptation of projects to local conditions;
- improved local skills;
- improved local awareness of project aims; and
- increased political support for the central government.

in each ministry for family planning, and a single officer in charge who is responsible for it. This amounts to having an individual and an organization pay attention to service delivery.

The first task of national planning is therefore to assure that its basic service delivery systems operate as they should. Before addressing issues of linking population, environment and development issues, they must be sure they can provide the most basic health, educational, and agricultural services.

2 Paul Harrison makes the same case in his excellent 1994 report to the European Commission, *Linking Population and Development*.

And they must be sure they can provide the physical infrastructure – roads, water, sewage and utilities – that a modern society requires. To meet many of these requirements, specialized service delivery capacities must be developed.[2]

When they are developed, then national planing can turn its attention to how they might be more effectively linked together. Then the problem of integration emerges.

Questions

It is important for a planning group to have a good sense of the strengths and weaknesses of the organizational tools that are to be used. A general sense can be gained of the overall strength of the political–administrative system with some relatively simple questions.

The simplest measure of the strength of the system lies in the capacity to deliver social services to the rural areas. Thus rural–urban differences in primary health care, educational enrolment, and family planning services will provide a crude but useful measure of how effective the system is in overcoming the barriers of distance to service people in rural areas.

Parallel to the rural–urban differences are rich–poor differences. For any service, from health and family planning to clean water, protected sewage, or utilities, one can ask how the service distribution differs for more wealthy and less wealthy groups.

These distributions are important for two reasons. First, sustainable development, or human wellbeing, requires that most or all people share in the condition. This is the basic point of *Caring for the Earth*. The great disparities between rural and urban areas, or between rich and poor, are both unstable and unsustainable. With large portions of people living in rural areas, for example, national infant

mortality cannot be brought down without substantial declines in these areas. A fertility limitation programme as part of promoting reproductive health will only be successful if rural people enjoy roughly the same access to services as do urban people.

In addition, the ability of government to provide services to rural areas and to the poor is a good measure of the general capacities of the political–administrative system. Governments that cannot serve rural areas often cannot do other things well either.

Measures of outcome such as these will not necessarily tell one why the system is weak or strong. Usually, weaknesses are found in staffing, financing and logistics. There can be many reasons, however, and most are quite location specific. Thus what planners need at this point are measures that indicate problem areas. From that point, specific observation and local knowledge must be used to determine the cause of the weakness and how it is to be overcome.

Another useful and readily available measure of the strength of a political–administrative system can be obtained from the UNDP which uses a Human Development Index, or HDI, to indicate development that is more than simply economic. Most useful here is the comparison of the HDI with national wealth, or per capita GNP. This indicates countries that have higher or lower levels of human development than expected from their level of wealth. Any country can see how it compares with others in what is, in effect, the capacity to turn wealth into human development. This capacity is not related to a country's level of wealth. Rich and poor countries alike may show less human development than expected by their levels of wealth; and rich and poor countries alike may show greater levels of human development than expected by their level of wealth. This comparison of countries should be

See Box, page 107, for a simple and effective measure of the effectiveness of a national family planning programme to deliver services to the rural areas.

illuminating to planners, since it leads them to ask why their country shows particular strength or weakness on this measure.

Three Arenas for Linking Policies

National governments can operate in three distinct arenas in integrating population with strategies for sustainable development. The three are partly related, but each also has its own set of demands and constraints, and will require the activities of specialized officers and personnel. At the same time, all three levels of action require national government leadership and policy formation. Thus they are all relevant for national strategies for sustainable development.

1. International

Participation, affirmation and ratification of international agreements and action plans. Inter alia, these include the Convention on Biodiversity; the Montreal (and subsequent) Protocols on CFC controls and CO_2 emissions, the Tropical Forest Plan of Action (TFAP), including proposals for Enhancing People's Participation in the TFAP; *Agenda 21*; and the Programme of Action adopted at the International Conference on Population and Development in Cairo in 1994.

> Participation in the international meetings gives voice to the country's special needs, aims, and objectives, ensuring that the global agreements reflect the specific conditions of the country.

> Acceptance of the agreements, including plans for implementation, give the country a role in the global community, both making it a responsible citizen of the community, and giving it a claim on the resources of the global community for its own actions to promote sustainable development.

The World Population Plan of Action has become controversial. It is important to recognize, however, that the areas of controversy in the Plan are smaller and narrower than the areas of agreement. While modern contraception and abortion are controversial, most other recommendations of the plan, such as increasing primary health care and education, especially for women and for the rural poor, find wide agreement among nations of the world.

2. National

In the national arena, government has responsibility for developing the specific plans and programmes that will promote sustainable development. This includes a wide range of activities from basic legislation to detailed planning and implementation, and the development of both structures and processes that mobilize local communities to work toward sustainable development.

3. Subnational

National governments have the capacity and responsibility to organize local units (states, provinces, districts, subdistricts etc) so that they will be active participants in the promotion of sustainable development. It is especially important that the national government decentralize control, to give local units a wide degree of latitude in adapting national plans to specific local conditions. This implies both devolution of authority, and allocation of resources.

See page 126, Measuring the Quality of Life.

See *Our People, Our Resources* (Barton et al, 1996) for strategies for local development and collaborative management.

Questions

At all three levels, the questions asked will be relatively simple, but the answers, and the analyses that generate those answers, will be complex.

At the international level we ask:

Did the country participate in the major international conferences: UNCED 1992, ICPD 1994, Social Summit 1995, Women's Summit 1995?

What steps has the country taken to carry out the programmes of action from each of these conferences?

At the national level we ask:

What policies does the country carry out that affect human wellbeing, especially for education and health; and for the wellbeing of the environment, including policies for environmental protection?

What subsidies affect natural resource use and environmental protection?

At the local level, we ask:

To what extent is authority and responsibility devolved to local units for promoting human welfare, and for controlling natural resources?

Integration

Integration is one of the most common terms found in plans and attempts to link different specialized functions. There have been global attempts to integrate population and development (1974 and 1984, and 1994), and the environment and development (UNCED, 1992). The ICPD Document (UN, 1995) also includes a statement on linking population, development and the environment. There have been countless national governmental attempts to integrate some combination of population, development, health, agriculture, the environment, education, and special issues of women,

and of indigenous peoples.

There are, however, a number of problems with integration. First, the term integration itself can have a number of meanings. Proposing integration still leaves open the question of how it should be organized. Secondly there is the distinction between statement and action. Merely mandating or directing that activities be integrated does not always lead to their integration in fact (see Box, page 31).

It is especially important to note first that there is no one best way to organize anything. There is not one best way to organize the linking of population with environmental issues in promoting sustainable development. The best way always depends on local conditions: political, cultural, organizational, and historical. There have been many attempts to integrate various specialized units of government, and it is possible to find both successes and failures in all types of integration. There are, however some useful generalizations that can be made. First, some definitional clarity is needed. There are two main types of integration: administrative and service.

Administrative Integration

Administrative integration involves placing a number of different specialized units under one administrative head. Environmental protection might be integrated with industrial policy by placing it in a ministry of commerce and industry. Population or a family planning programme might be integrated with health by placing it in a ministry of health. This type of structural change can be useful when a weak or especially important activity needs the protection of a powerful organizational unit. Development units are typically placed close to the centre of political power to give development the support and power that comes from the political centre. In the early 1980s, for example, Pakistan's weak family planing programme gained considerable support by being transferred from the Ministry of Health to the much more powerful Ministry of Planning. (That the support did not last was due to other political and administrative problems). Often, however, the proposal for this type of administrative change stems from a desire simply to enhance the power and resource claims of a particular unit. Moreover, it is in this type of organizational change that the difference between statements and actions become most critical (see Box, page 33). Administrative integration is well known to enhance the position of a particular ministry or minister, but it is not well known for achieving an effective linking of specialized services.

Service Integration

This involves two or more specialized services working together at the actual point of service delivery. Agricultural extension services and experiment stations remain separate organizational units, but their workers meet together in the farmer's field to work together on increasing yields or controlling pests and diseases. This type of integration is generally associated with successful service delivery to the user, though it requires some linking mechanism to facilitate communication and scheduling.

There is another lesson to be learned about the structures of integration or service delivery. A structure cannot be effective alone; all structures depend on the work that is put into them. A good illustration of this is available from a comparison of the Indonesia and Philippines family planning programmes. Both have very much the same structure, with a major top level coordinating board designed to bring together all the units whose cooperation

---

Integration: What Works Depends on Local Conditions

A study of family planning programme performance in Korea provides useful lessons for integrating strategies. The study measured integration with a simple question to family planning field workers: 'How much time do you give away to other units and their services?' The question was based on the recognition that county chiefs often directed workers to assist one another in such things as health campaigns, agricultural projects or even tax collection. The study also measured the effectiveness of each family planning programme by the simple ratio of acceptors to workers.

Overall, there was no relation between the amount of time reported 'given away' and county level programme effectiveness. This was because the relationships were reversed in rural and urban countries. In rural areas, the more reported time given away to other activities, the higher the effectiveness of the county family planning programme. In urban areas, the reverse was true: time given away reduced programme effectiveness. Researchers asked why.

The answer lay in the differences between rural and urban conditions. In rural areas, working together implied a more effective team spirit. Family planning workers helping other government officers would bring return assistance to them from an effective rural county government team. In urban areas, teamwork was less important because of the more anonymous character of urban clients. Family planning workers were more effective when they were like energetic sales people, making more contacts to more prospective customers. Taking time away from family planning took time away from their own 'sales' efforts, thus reduced the effectiveness of the clinic in recruiting acceptors.

Source: ESCAP 1975

---

is necessary for the success of the programme. In Indonesia the structure works very well, producing a programme that is widely regarded as producing excellent results under very difficult circumstances. Overall contraceptive use is about 50 percent, and the total fertility rate has dropped from 7 to 2.9 percent. The same structure in the Philippines has produced far less. Contraceptive use is only about 40 percent, and the total fertility rate has declined far less, from 7 to 4.1 percent. Indonesian leadership put a great deal more effort into their structure. Structure alone will not create integration.

Integration by Local Tailoring of Structure and Process

Actually achieving effective integration of disparate units, agencies and activities is difficult, in part due to the natural resistances of specialized units. But it can be done, and there are many successes to help guide us (see Box page 35). At any level, people from different agencies can be brought together in committees, task forces, or working groups. The publication *Strategies for National Sustainable Development* (Carew-Reid, 1994) provides many examples. There are, however, three useful lessons to be learned from experience with integration.

1. Participation: small elite task forces seldom achieve much success. Broad participation, on the other hand, is more successful.
2. Local adjustment: tailoring strategies to local circumstances is also important. Plans cannot be developed from the outside and dropped into a country. They are best developed locally, where they will benefit from knowledge and sensitivity to local political, cultural and ecological conditions.
3. Sustained action: success also requires sustained action, which implies political will, or top level political support in order to continue putting resources and work into the structure.

Questions

Asking questions about integration involves three steps.

First, what activities does one wish to integrate, and why? It is most important to be clear about the why. Usually arguments are made that integration will reduce costs and improve the delivery of services. If such reasons underlie the decision to integrate, they should be made explicit.

Second, if integration has been mandated, it is important to ask to what extent the mandate is carried out. To what extent are the different activities actually brought together? Note the case of mandated integration in Papua New Guinea in the Box below (page 33). Integration was directed, but did not actually occur in service delivery.

Third, the effects of the integration must be assessed. Is there improvement in all the services brought together? Or do some improve at the expense of others? Or do all suffer from the new patterns of integration?

Participation and Benefits

USAID (1994) identifies six important lessons learned from its Integrated Conservation and Development Projects (ICDP). These are directly applicable to the problems of linking population with both development and conservation. Although these lessons do not address the organizational problem directly, they provide important guides as to what the organization of integration should produce.

1. Material benefits from conservation must be real and clearly linked to conservation. Conservation can provide material benefits, and must do so in order to be acceptable. If conservation reduces a peoples' material benefits and must be

subsidized from outside, it will not be sustainable.

2. Project benefits must be equitably shared. Stakeholder analysis – or identifying which individuals or groups have a specific interest or stake in an area – should be used in planning. Too many development or conservation projects fail because they bring benefits only to the wealthy or to those outside the system. Local participants can be the best protectors of the environment, but to gain their assistance, they must be included in the benefits and not just in the work.

3. Stewardship and ownership are vital. To assure that costs and benefits of a project are shared equally, local residents or participants must have control over basic resources. This implies the need for an effective form of ownership or rights of use.

4. Incorporate local knowledge. Local participants are the ones with the best knowledge of how the local ecosystem and culture are connected together and how they work. It is essential to draw on this knowledge to achieve successes in linking environment and development.

5. Give attention to the policy environment. What is possible locally often depends on the larger political environment, and the policies adopted by the political centre. A project to link environment and development must be sensitive to the central policy and, where appropriate, attempt to influence that policy.

6. Consider both biological and socioeconomic criteria when selecting or designing projects. The technical expertise required in selecting areas and designing projects must include both the natural and social sciences. Drawing these together is not easy, but it is imperative for success.

Failures and Successes in Integration

A Failure of Administrative Integration: When Papua New Guinea launched its family planning programme in about 1980, it (administratively) integrated family planing with maternal and child health services. Primary health care nurses were directed and trained to deliver both types of services in both rural and urban clinics. This substantially increased the budget and personnel of the MCH unit, with foreign donors increasing their contribution specifically for the integrated service. But in most primary health care clinics, the nurse scheduled different times for MCH and for FP services, and failed to integrate the actual services. In the family planning clinics, there was no mention of or instruction in maternal and child health, nor were there vaccination services. Similarly, in the MCH clinics, where vaccination was scheduled, there was no mention of the health aspects of birth spacing, breastfeeding, or of limiting births among very young or older women. Thus the potential benefits of the integration were felt mostly at the upper levels of the administration, but were not passed on to the clients at the point of service delivery. No increases in contraceptive use or reductions in fertility were recorded, nor was there much improvement in the levels of infant or maternal mortality.

LESSON: Integration is not achieved simply by mandating it.

A Success of Service Integration: When Malaysia began its national family planning programme in 1967, it created a separate Board to direct the programme, keeping it administratively separate from the Health ministry. Each maintained its own budget, personnel, goals and plans of action. But part of the FP plan was to send its workers to the MCH clinics to provide family planning instruction to women who brought their children in for MCH services. The family planning workers went to the regular MCH clinics in the towns, and travelled with the MCH workers in the mobile van that served more remote rural areas. Communication and friendships developed among the two sets of workers and women attending the MCH clinics received good instruction in both MCH and family planning. Contraceptive use rose and fertility fell, and both infant and maternal health improved measurably.

LESSON: Integrating services of different agencies at the point of service delivery worked well.

Organizational Options

1. Place Population Specialists in Specialized Development and Environmental Agencies.

Population specialists go by many different names. The core discipline and scientific technology is that of demography, or the study of the size, territorial distribution and composition of a population, as well as its changes and the components of those changes. The technology of demographic analysis is highly developed, and exceptionally powerful. It can be used effectively in a wide range of analyses, including policy planning, programme implementation, and evaluation and monitoring. Demographers or demographic skills and training are found in many of the social sciences – including sociology, economics, political science, history and geography, as well as in business studies. Population specialists are also trained in public health, where demographic tools are part of the package that also includes biostatistics and epidemiology. In addition, training in public health usually includes skills for analyzing policy formation, programme implementation, and the evaluation of policy and programme impacts. Over the past four decades, international training programmes have produced a large corps of population specialists, working in government agencies and universities in all parts of the world.

Moreover, demographic techniques – measurements, data and models – have been extensively computerized, and fitted with clear and attractive graphic displays. In addition, there has been a rapid growth in basic population measures, size, birth and death rates, age–sex compositions and territorial distributions. While many data collection systems are deficient, the tools of demography also provide capacities to estimate errors or gaps, and to make accurate estimates from limited data.

Thus the population specialist is a bureaucratically portable tool, like an economist or a medical doctor. If one wishes to include population issues in any specialized activity, such as agriculture, education, or environmental protection, one of the simplest ways to do this is to recruit a population specialist.

Advantages: Organizationally the easiest to accomplish; provides a powerful and useful specialized tool of particular relevance to the connection between population, development and the environment.

Disadvantages: No assurance that the tools will be used, that effective communication will take place between the demographer and other specialists; no necessary link to policy planning and implementation.

2. Create a Specialized Development–Environment–Population (DEP) Team for a Specific Ministry

Identify a team of specialists in development, environment, and population. Assign them the task of reviewing critical development–environment–population linkages and proposing specific areas for planning, policy making, and for project interventions that will provide some leverage toward promoting sustainable development. This can be done in any sectoral agency, such as agriculture, health or forestry, or in the central planning agency. It can be done at national level, or at a lower, state, province or district level.

Advantages: A small team is relatively easy to identify and staff with people who are compatible and willing to understand another discipline. It would be able to review its specific sector and point to areas where critical policy changes or programmatic interventions could be useful.

Success in Inter-Agency Cooperation: Malaysia's Rural Development Programme

From 1957 to 1968 Malaysia mounted an extremely successful rural development programme, with heavy emphasis on physical infrastructure construction. Strong central leadership generated high levels of local participation at state and district level. Projects were initiated from national to district levels, with the active involvement of all government offices as well as local, state and national elected members of government.

Structure: The core of the action was public construction, directed by a new Ministry of Rural Development, directed by the Deputy Prime Minister. A new structure was created. Rural Development Committees were established at national, state and district levels. The committees included all members of government technical departments and elected leaders. They were instructed to meet regularly to plan for a series of needed public works in their areas. A large (one metre square) Red Book was created with maps of the district, state or nation. Map overlays showed existing and planned projects, and wall charts located each project with a schedule of the steps necessary to complete it. Weekly entries were made on the charts to indicate whether they were on, behind or ahead of schedule.

Process: Into this structure, the Minister instituted a periodic briefing process. State and district committees conducted weekly meetings, often including a briefing for the Minister or one of his staff. Briefings included all relevant government officers, expected to be able to explain in detail how each of the projects was going, and where there were problems. The briefing was the point at which the normal government inter-agency process was open to scrutiny, providing accurate information on issues of accountability. With wall charts showing progress, and all relevant government officers in the room, it could readily be known who was responsible for delays or for especially notable achievements. The existence of the Minister at these briefings meant that power and authority were introduced into the process. Rewards and punishments, based on accurate information, could be meted out on the spot, and they were.

Outcomes: Everywhere in the country roads, health centres, schools, water and electricity services, small scale irrigation schemes, mosques and government offices were being built. Land was being opened and houses and farms created for previously landless farmers. The level of construction in the country was widespread, highly visible, and did a great deal to increase the quality of life. Among government officers, there was both fear of the Minister's power and confidence in its fair use. There was also a high degree of morale, based on the knowledge that if one did one's job, progress could be seen, and upper level restrictions, delays or rejections could easily be overcome. The people repaid the government with a substantial electoral victory in 1964.

The Importance of Leadership: The system required strong leadership. From the start it was provided by the Deputy Prime Minister, Tun Abdul Razak, well known for his administrative skills. In 1963–4 when Indonesia launched its 'Konfrontasi' attacks on Malaysia, Tun Razak, who was also Minister of Defence, was forced to shift his major time, energy and emphasis to defence. The Rural Development Committees continued to meet, but meetings became less frequent and less critical. Processes of inter-agency communication became less lively and more routine and agencies went along on their own. Achievements dropped off as government put less strong leadership into the process. In the next election, the people rewarded the government's backsliding with a near defeat that triggered serious race riots.

Source: Ness, 1964.

Over time, the team could build a repertoire of lessons and a reputation for useful inputs that would bring development, environment and population thinking into close integration throughout the upper levels of government.

Disadvantages: A small team can deal with only a small number of conditions, locations, or observations. It may miss important environmental impacts. Unless it is properly placed, near the centre of power and influence, its analyses and suggestions may get no action.

> *3. Create a Cabinet Level Structure and Process for Bringing the Three Areas Together in Periodic Reviews of Long Term Development Plans*

This is perhaps the most common suggestion for integrating anything with anything else. It is common because the integrators recognize the importance of top level commitment and leadership, or 'political will', to make integration work. Currently much economic development planning is done with some form of top level unit that acts as an executive committee, and brings together the various agencies whose active cooperation is needed to turn development plans into reality. A cabinet level task force for development–environment–population (DEP) can be created as a subunit of the national planning committee. It is important not only that there be a DEP task force, but that it has a regular schedule of activities, such as weekly or monthly meetings, and a process of systematic monitoring and evaluation by which the various units of government can be held accountable for their actions in achieving the common goal (see Box, page 35).

Advantages: Brings together all specialized agencies of government. Gets relevant information into the overall national planning process. Increases the probability of quick and effective action to deal with, for example, environmental problems. Builds into all specialized agencies the capacity to communicate and cooperate across agency boundaries. Is capable of setting goals and assigning responsibility for goal achievement. Capable of clearing bottlenecks that come from inter-agency unwillingness or inability to communicate.

Disadvantages: Difficult to put into operation. Requires strong and committed leadership. Requires high quality information for the exercise of leadership. Depends highly on structure and process (see Box, page 35).

Questions

The questions to be asked in considering organizational options are essentially questions about where specific technical capacities exist, and the extent to which they are being used. On the population side, the technical capacities are relatively easy to identify. People trained in demography, in public health or in population planning are needed. Demographic training is now part of many different social science disciplines. We find anthropologists, economists, geographers, political scientists, psychologists and sociologists with demographic specializations.

On the environment side it is less easy to identify the needed specialization, because there are so many. Ecology is the most general field, but many other specializations are also relevant. The specific type of specialization required should be determined by the unit in which the linking is to take place. For example, a forestry agency could be the base, implying a population specialist working with foresters. For the third option listed above, a cabinet level unit, one would need a range of environmental specializations, from ecology to hydrology to public health and toxicology, as well as a variety of fields such as forestry and agriculture.

GIS: New Tools for Linking Visual Capacities

One of the major advantages and disadvantages of specialists lies in the diagnostic tools they develop. All specializations have special tools for observation. Some of these are clearly identifiable physical instruments. They range from the microscope, the telescope and the scale to a wide range of instruments that can sense various bands of radiation. These are physical tools whose increasing refinement and sophistication permit scientists to increase immensely their powers of observation. In the social sciences tools are also developed by specializations, but they often have a different character. They tend to be more conceptual than physical. GNP, for example, is a concept of the wealth of a nation. It is measured through the collection of masses of economic data on the value of goods and services. This instrument permits us to see the wealth of a nation and to ask questions about the source and the use of that wealth. Sociologists use large scale probability surveys to observe behaviour, knowledge and attitudes, all of which tend to be classified by some concept of social organization. Psychologists use experiments on humans and other animals to observe how physiological and intellectual processes work.

Each discipline develops increasingly sophisticated tools for observation. This increases the power to see, but it also reduces the capacity to communicate those observations and perceptions to others who lack the highly developed skills needed to use the instrument. Thus in linking different specializations, either scientific or governmental, great advantages are to be gained from developing tools that permit different specializations to share their powers

of observation.

Recently scientists from many different disciplines have been developing tools that permit us to see the area or geographic distribution of many things, from land use to human behaviour. This new set of tools of observation is generally called Geographic Information Systems, or GIS. It began with simple mapping, and is now greatly enhanced by electronic sensors that permit the generation of digital maps. These allow us to see directly the distribution of any specific condition, and they allow us to represent that distribution quantitatively for powerful statistical analysis and manipulation. This means that just as demographers can project population movements into the future using different assumptions that produce difference scenarios, the same can be done for such things as land use, water flows, and human impacts on the environment. This development greatly increases our capacities to link population and environmental issues in promoting sustainable development.

The Malaysian 'Red Book' (see Box, page 35) provides a good example of the use of maps and overlays to visualize a process over which one wishes to establish control. In that case, it was used primarily to evaluate, oversee, and push along the process of public infrastructure construction. The same tool can be used to manage population–environment planning as well. Moreover, the tool has the special capacity to help locate what can be called hot spots in the DEP nexus: a wetland scheduled for a housing development, which could better be used for a natural waste management plant, for example, or a forest threatened by immigration and cutting which would destroy a watershed and lead to flooding. Interventions can then be planned and activated for those spots.

Maps and overlays can be used at the national level, or at any subnational level, even down to small villages, though we shall leave those smallest units for the companion volume on local communities, *Our People, Our Resources* (Barton et al, 1996). The process involves a basic political map, showing administrative boundaries, roads and major built up areas, with transparent sheets that can be overlaid on the map. Each of the transparent sheets is used to indicate specific conditions relevant to planning. There may be, for example, one or more population maps, showing major population concentrations with numbers, density, and possibly other relevant socio-economic characteristics. One map can show the basic population concentrations, another the age distribution, or the industrial distribution of the labour force, and special populations such as indigenous peoples. Such maps can also be used to show the incidence of diseases, or poverty, and all can show changes over time, both for the past and projected for the future. In addition, there may be maps showing major environmental characteristics, especially those relevant for nature conservation and sustainable development. Forests, vulnerable species, wetlands, marine systems, lakes and waterways and their pollution levels, wild species censuses, agricultural land, patterns of erosion or land degradation all would be useful for seeing the juxtaposition of population and environmental conditions.

Finally, major development projects and plans may have special overlays. These overlays can be used especially to identify areas where high population densities or development activities, or both, impinge on fragile or vulnerable ecosystems. These might be called DEP hotspots. For these areas, specific studies and relevant schemes of intervention can be undertaken.

The maps and overlays described above are often included in GIS, a powerful and rapidly growing technology with powerful computer applications and attractive graphic displays. Computer technology has grown very rapidly, and now even permits one to portray relationships that are almost impossible to see with the unaided eye on any type of map. For example, early detection of crop diseases or underground water seepage patterns can be mapped and related to a variety of natural and human conditions, such as land forms and industrial locations.

One especially useful potential emerging from GIS is the capacity to use remotely sensed data in computerized analyses. Remote sensing can be done from aeroplane fly-overs, but massive coverage is available from satellite imagery, which is becoming both more refined and cheaper.

The advantages of the new electronic imagery are immense. The fact that images are electronic, rather than chemical (photographic) means that they can be computer enhanced. Thus different colours can be used to indicate very small differences in the wave lengths of the spectrum being sensed. This makes it possible, for example, to see conditions, such as plant diseases, before they become visible to the naked eye.

The electronic imagery also makes it possible to distinguish built up areas from vegetation, and thus to measure precisely the number of hectares of an urban area. Images from different time periods can also be overlaid, with distinctive colours used to show those specific areas whose land use has changed from one time to the next. Rates of deforestation, desertification, or urbanization can thus be measured with a high degree of accuracy.

Many countries already extensively use satellite remote sensing and GIS in their planning. Few, however, are yet putting population, environmental and developmental data into the same analysis.

Questions

There are two types of questions to be asked in considering these new tools of observation. One concerns skills, or human capital, and the other concerns physical tools, especially computer capacities and satellite imagery.

First, do we have people with training in GIS? This is now a well recognized specialization, making it relatively easy to identify available skills, or to obtain relevant training for people who will undertake the activity. It is important, however, to recognize that GIS is not simply a map-making activity. Too often expensive equipment is simply used to make maps, indicating that the training has been limited to cartography and has not included the broader range of GIS skills. Those skills basically train an individual to provide a wide range of information on human activities and environmental conditions and to integrate these in their spatial distribution.

Second we must ask what equipment and data are available. Computers are now used to manage maps and to display all manner of data on computerized maps. Programs for such data management vary greatly. One group has produced a program called QUICKMAP, which uses less than 100k of computer memory, and enables the user to create digitized maps and to display a wide range of data simply on the maps. There are also very large and expensive systems that have great power. The equipment to be chosen depends on the uses to which it is to be put, and such decisions should be made locally by qualified people.

In addition to questions about computer equipment, we should ask what kinds of maps and data are available. Here the most rapidly growing information comes from satellite

Geographic Units for Integrated Planning

Planning and implementing development, environmental or population projects is typically done by administrative areas: nations, states or provinces, districts or countries, towns or municipalities etc. This is proper and natural, since governments count people, collect taxes and allocate authority, responsibility and goods and services by such administrative units. It raises problems, however, when administrative districts cut across natural ecosystems, such as watersheds, wetlands, forests, or deserts. This means that different political or administrative units share responsibility for a given natural system. The effective management and protection of that natural system will thus require cooperative planning and programming by more than one unit, which is often difficult to achieve.

This problem has typically been solved, often very effectively, by regional groupings. The Tennessee Valley Authority (TVA) developed in the US in the 1930s is a model of effective regional development, though its mandate is narrower than those that would be developed today for integrated planning for sustainable development. As with many aspects of this planning, the TVA experience shows that creating an appropriate ecosystem planning area requires the intervention and strong leadership of an upper level political unit.

It is also problematic, though often less so, when more than one such natural system is included in the same administrative boundary. Here the problem is to manage the individual natural systems not as single discrete units, but as a system of such units. In the case of national parks and protected areas, for example, there needs to be a national level planning capacity that can see all of the parks and protected areas as a system of parks, linked together geographically by possibly unprotected areas. If the parks are small and scattered and the unprotected areas are large, even extensive protected areas in total acreage may not be sufficient to promote the fundamental biodiversity that is required for long term sustainable development.

imagery. They are in electronic form and thus should admit extensive computer manipulation. Many countries and private or semi private organizations today produce satellite imagery and make available either the data on computer tapes, or images with data tailored to the needs of the user.

Population–Environment Networks: A Proposed Strategy

Bases for Action: Lessons Learned

IUCN policy evolution and the past programme activities have provided a wealth of information about population–environment relationships. The major observations and lessons derived from these case studies are as follows.

1. Location specificity

IUCN field projects show a great variety of population–environment relationships. One can find evidence of population dynamics associated with environmental enhancement and protection, or with environmental degradation. There is no single, simple population–environment relationship. All relationships are location specific.

Thus linking population and environment must be done in specific locations, by people with good knowledge of local conditions.

2. National planning for sustainability does not always give full consideration to population issues

Plans sometimes lack technical inputs needed to produce adequate consideration of the full range of population conditions and their impact on the environment.

Thus national plans must be built on the full range of population and environmental expertise.

3. Local implementation is needed

National plans need to be implemented at local levels. This requires both enabling activities, and assistance that respects and pays attention to local knowledge and interests.

Thus local population and environmental specialists must be brought together to work on specific local problems.

4. Expertise is available

In all IUCN field locations there are well trained and experienced population specialists, including demographers, social scientists and public health practitioners. Too often these human resources are not effectively utilized in planning for sustainable development. Specialization keeps people apart; bridges are required to bring different specializations together.

Thus population and environmental issues can best be integrated by developing local networks that bring population specialists into close and on-going relationships with IUCN staff, members and partners – the wide array of people who work in environmental conservation.

5. Local communities often provide the best setting for dealing with population–environment relationships when they are assisted to identify and promote their own needs.

Conservation and development efforts often fail because they ignore local communities and do not gain their participation.

In addition to national level planning and implementation, it is important to develop mechanisms by which local communities can be assisted to address their own wide range of closely interconnected problems.

These lessons lead us to propose a strategy of building Population–Environment Networks (PENs) to create practical ways to link population and environmental issues in national and local planning.

Although PENs will differ with their location, and must be built at the national level on the basis of a good knowledge of existing local conditions, we can propose a general plan which could provide a framework to be adapted to distinctive local conditions.

Building PENs

First, the decision to establish a PEN must be taken by specific groups and agencies, and an organizational home must be established. This can be a government agency or a non-governmental organization, or a university. The decision must be made locally on the basis of judgements of the flexibility and support that can be provided by either government or non-governmental organizations. When that decision is made, the work of building a PEN will include four specific steps and activities: establishing, activating, supporting, and evaluating the PEN.

Establishing a PEN

To establish a workable PEN it is necessary to identify local population specialists and to bring them together with environmental specialists. To provide both technical and financial support, it will be useful to establish some kind of advisory committee of government, non-governmental organization and

donor agency personnel. We can suggest the following five elements of establishing a PEN.

1. Steering Committees

Establish contacts with relevant government agencies, donor agencies such as local bilateral aid agencies and UN offices, and non-governmental organizations such as the International Planned Parenthood Federation (IPPF) and its affiliates. Add to this representatives from the most relevant IUCN members and partners to identify appropriate environmental groups. From these potential population and environment organizations, create a small steering committee. The committee will provide ideas, advice and contacts to bring together the specialized organizations working in population and environmental activities. It will also provide assistance and advice in later stages of activating the PEN, when it will be necessary to find funding for project proposals.

2. Population specialists

Identify population specialists (demographers, social scientists, public health officials etc) in the country. Use personal connections as well as lists from national and international organizations, such as the UNFPA, International Union for the Scientific Study of Population (IUSSP), or Population Association of America (PAA). Solicit professionals asking them to express their interest in becoming part of the new networks through a formal process of invitation; they may then be entered into a central data file with information on professional background and experience.

3. Environmental specialists

Identify appropriate environmental specialists. Use personal contacts with IUCN staff and other IUCN members and partners. As with the population specialists, solicit them for expressions of interest in participating, and their responses will become part of the central participants data file.

4. Projects

Identify projects currently undertaken by IUCN, environmental groups, and by local population groups that are potentially relevant for linking population and environmental conditions. These can be used as cases to be addressed in workshop exercises. They might include projects in primary environmental care (PEC), or protected areas at the local level, which do not presently have a distinct population component. They might also include national level sustainable development programmes that are weak in population elements. Alternatively they may be local or national population projects that operate in vulnerable environments and that lack an appropriate capacity to address environmental issues.

5. Workshops

The population specialists and environmental specialists identified in activities 2 and 3 can be brought together in one or more workshops. The aim of the workshops will be to establish communication and capacities for collaboration between the two groups of specialists. The workshops can take participants through a series of exercises, drawing on their own field work, experience and expertise, by which the two sets of professionals can share their perspectives, theories and technical tools. The task should be to have the participants present exercises to one another, so that they are teaching one another the distinctive skills and experiences they have. But there should also be specific exercises that bring new teams to work

together to solve problems. The specific number and location of the workshops, as well as their size, however, must be decided upon according to local opportunities and constraints.

The specific content of the workshops must be developed with local data and experience for each country location. The extent of forests, deserts and wetlands, the amount and quality of water, and the conditions of the population will dictate what specific data should be used and which questions should be asked in the workshop. Nonetheless, it is possible to suggest a list of data and issues that can be used effectively in most cases. These include:

a. Population data, as discussed above, including size, distribution, age and sex compositions, and rates of birth, death, growth and migration.
b. Environmental problems including land use change, expansion of agriculture and changes in agricultural yields, soil conditions including erosion, changes in forest cover, water availability and changes in quality and quantity, and changing levels of pollution in land, air and water.
c. Population–environment linkage questions which might include the following.

How are land-use changes linked to population dynamics? Is the nutritional level sufficient to support good health for the population? To what extent are population growth and changes in family income affecting the demand for food? To what extent do population growth from natural increase and from migration affect the demand for land, the rate of deforestation, and the destruction of species habitat? To what extent do population growth and patterns of consumption and production affect levels of air, water and land pollution?

In all cases where population is considered, questions should be asked about how the composition of the population – by age, sex, education, and income – affects the impact on the environment. Conversely the same questions can be asked in reverse; how do different patterns of environmental change affect people of different ages, sex and wealth?

Below we present a series of hypothetical scenarios of what the outcomes of such questions might be.

This creates the network, by bringing specialists together and getting them to explore how they can work together. The network must then be activated – put to work to build up some momentum.

Activating the PENs

One way to activate the network involves providing financial and technical support for local projects proposed by PEN participants. This plan for activating the Networks is based on two premises. One is that a network functions when people in the network work on real problems. Thus we need to generate teams of people doing things together, identifying population–environment problems, and ways to assess those problems or ways to produce interventions that will work. The second premise is that the most important, and actionable, population–environment relationships are location specific. Designing projects that identify critical linkages and tactics for interventions to promote sustainability must be done in specific locations. Such interventions may include activities to reduce population pressures by affecting migration streams, reducing growth rates; or they might be activities to promote environmental protection or sustainable use of resources.

1. Supporting projects

To activate a PEN, a fund for pilot projects could be established and managed by a local selection committee, which may also include members of the PEN steering committee. The funding committee would call for project proposals from PEN participants, and offer support to initiate population–environment projects that participants find are most relevant for their situations. These may be national planning projects, or local community level projects, utilizing the PEC approach that IUCN has adopted.

2. Monitoring and evaluation

Logical Framework Analysis could be promoted as a tool for project formation and evaluation. Projects would thus have clearly stated objectives, whose achievements could be marked by observable indicators. This will provide clear, objective indicators of performance for each project. Annual reviews could be made of projects to evaluate their impact on population and environmental conditions.

3. Duration

One possibility would be to provide financial support for pilot projects for five years. By the end of that time, it could be assumed that successful PENs will have developed sufficient momentum to be self-sustaining. Integrated activities will have proven themselves of sufficient utility to be built into public and private development or conservation programmes, and the projects undertaken by PENs participants will have proven sufficiently productive to be able to attract public and private funding for their continuation. Moreover, PEN participants would have the funds and sufficient personal and professional rewards from the collaboration to lead them to continue it.

Supporting the PENs

A number of things can be done to support the PEN as it is in operation. These will include activities to identify participants and to establish communication processes by which the lessons of individual projects can be brought to a broader audience. Four specific activities can be suggested here, but it is assumed that each network would develop its own repertoire of activities relevant to the specific location.

1. A roster should be maintained of active and prospective network members.
2. A periodic newsletter could be published and circulated providing information on the pilot projects and the lessons learned.
3. Case studies could be undertaken and published providing details on the life of projects, indicating causes of both success

and failure in managing local population–environment linkages.
4. Periodic workshops could be held to draw new members into the workshops, to present case studies, and to identify and codify the lessons being learned from the pilot projects.

Evaluating the PEN

The PENs initiative will require a special form of evaluation, since it involves a number of different steps, and the specific content of the most important step – Population–environment Project contents – cannot be precisely determined in advance. We can, however, illustrate how that step will be evaluated by providing some hypothetical case scenarios, developed from existing IUCN activities on the ground at this time.

Establishing the PEN

Establishing the PENs involves identifying population and environment specialists and bringing them together in a workshop for mutual training. Two steps of evaluation should be undertaken.

1. *Participants.* The first line of evaluation is the number, position, and quality of the participants, or the people who indicated a willingness to be included and have committed their time and energy to the initial workshop. Three questions are asked:
   (i) Numbers. Are there sufficient numbers to produce an active network? Although exact numerical standards are not available, we should expect 30–40 participants to provide a critical mass to make the networks viable.
   (ii) Diversity. Do the participants represent the full range of specializations needed?

Steps in the evaluation process

I. Establishing the PEN.

Evaluation   A. Number, position, and quality of participants enrolled.
             B. Workshop evaluated by participants at its conclusion.

II. Activating the PEN.

Evaluation   A. Number and quality (judged by selection panel) of project proposals submitted.
             B. Project outcomes: impacts on population and environment (see hypothetical scenarios).

III. Supporting the PEN.

Evaluation   A. Number and quality of newsletters, PEN meetings and case study documentation.

There should be demographers, social scientists, public health and family planning professionals, and conservationists representing various areas including forests, wildlife, and wetland and marine environments, etc. There should also be representatives of government and NGOs, as well as university professors and scientists. It is also important that participants come from all levels: from field workers on the ground working directly on a day-to-day basis with communities, to academics and government officials.
(iii) Quality. Do the participants represent a high quality of the specializations and organizations from which they are drawn? The PEN should include both well known and respected senior members, and younger members recognized for their professional promise.

These questions could be addressed by the PEN Steering Committee, and provided in an initial report after the project has been running for about six months.

It should be noted that, although the recruitment of good participants is not a sufficient condition for the success of the PEN, it is absolutely necessary. If the PEN initiative cannot attract adequate numbers of sufficiently diverse and high quality people for the participants, it will not be able to act as an effective network.

2. *Workshop.* The second step involves the evaluation of the initial workshop. The workshop can be evaluated by an anonymous questionnaire distributed at the close of the workshop. (It is likely that the organizers will also hold an evaluation discussion session halfway through to determine how the participants judge the course of the workshop, and which

portions should be strengthened or dropped.) This will seek the participants' candid appraisal of the process, the mutual teaching and learning, the quality of the organization and leadership, and of the quality of the exercises prepared by the PEN office for the workshop.

Again, although a successful workshop is not sufficient to make the initiative succeed, it is vital. Specialists must be brought together to teach and learn from one another so that they can define practical problems that they can address as a team. Without developing the capacity of PEN participants to talk and work together, the network cannot be functional.

3. *Schedule.* This evaluation could be made at the end of the initial workshop, probably during month six of the initiative.

Activating the PEN

The PEN Funding Committee should develop procedures for calling for and funding proposals from the PEN participants. This will include announcing competitions with guidelines and rules, organizing a panel of peer reviewers, and developing guidelines for the Steering Committee to use in evaluating and deciding upon specific projects to be funded.

Information on project funding can be provided to the participants during the workshop. This will include the process of developing and submitting proposals and the criteria for their selection. While specific criteria must be developed locally in each PEN office, requirements would normally include a multi-discipline and multi-organizational team, and activities that integrate population and environment issues effectively in addition to having an identifiable impact on both population and environmental conditions. Proposals will also have to specify how they are to be evaluated. Again, two steps can be envisaged in the evaluation of the projects.

1. *Number and quality.* First, performance of the initiative can be judged by the number of proposals. Again, this is a vital condition, though not the only one. If no proposals come forth from the participants, the initiative can be considered to have failed early in its development. While exact numbers cannot be specified, one could expect five to ten proposals emerging from the workshop and the first announcement.

    The quality of these proposals will be judged by the steering committee making the selection for funding. One could expect two or three proposals to be sufficiently well developed and of sufficient importance to warrant funding. Peer reviewers and steering committee members will assess the quality.

    Schedule. This evaluation can take place during months seven and eight, after the initial workshop, announcement and selection of projects for funding has been completed.

2. *Project impact.* This is the most important step in the PENs initiative, and also the most important for the evaluation. It is also, however, the most difficult. The difficulty derives from two conditions. One is typical to all development projects. Such projects are designed to enhance human produc- tivity or welfare. It is never easy to assess such progress. It is even more difficult to determine what caused the progress and what impact the project itself had.

    There is another problem with the PEN initiative, however. While we expect

PEN projects to promote the wellbeing of people and the environment, we cannot say in advance what those projects will be. In fact, since a major aim of the initiative is to build capacity, the projects must not be specified in advance by any external agency. Moreover, all we know about population environment dynamics says that the most important dynamics are location specific. Thus it is local population and environment people who must identify problem areas and possibilities for intervention that are distinctive to their local (national to community level) conditions.

While we cannot (and should not) specify project contents in advance, we can provide a series of hypothetical case scenarios showing the kinds of projects that could be expected, what their impacts might be and how they will be evaluated. The scenarios are not pure fantasies, however. They are based on what we have seen in the field, and what IUCN staff are actually doing there.

Supporting the PEN

As noted above, a series of activities to support the PEN and to link its members to other relevant national and international activities should be undertaken at an early stage. One of the most important steps is to document the activities of the participants, especially in the projects that are funded under the trial period. This could be done in two ways. First a bi- monthly newsletter could be established to give participants information of the project proposal process, and the specific projects selected for funding. The newsletter could also provide brief notes on other IUCN projects, and on news of donor interests in supporting various projects. Each country newsletter could also carry information on initiatives in other countries, and of IUCN headquarters activities in population and environment matters.

In addition, participants could be invited to (for example) quarterly meetings hosted by the PEN office to hear of the progress on field projects and other news relevant to the participants. Personnel from international agencies and bilateral donors could also be invited to these meetings, giving PEN participants an opportunity to establish contacts with donors who might be interested in funding their projects.

Finally, the progress of the field projects should be documented, based on site visits and writing brief notes of the cases. These can be accumulated into case studies of the specific sites of the projects. They can also be included in the monthly newsletter, and accumulated at IUCN headquarters, together with those from other countries, to provide materials for a bi-monthly newsletter on population–environment dynamics.

Evaluation. The number and content of the newsletters would provide a partial quantitative indication of the activity, as would the number of meetings and attendance at the meetings. These are, of course, only means to the overall aim of the PEN – promoting the welfare of people and ecosystems. They are indicators that can easily be quantified and reported, but it should be noted that they are not ends in themselves. That objective – the joint promotion of human and ecosystem welfare – requires other measures, such as those we have seen above in discussing the impact of PEN projects. It is never clear in project evaluation just what the linkages are between the various elements of a project, and especially between those elements and the overall aim. We believe that communication processes like newsletters and meetings are important to support the PEN, especially since what we are creating is new

networks that take people from existing demands and link them to people and activities normally outside their range of communication. As with other aspects of this project, we believe these communication tactics in support of the PENs are necessary but not sufficient for the success of the overall project.

Scenarios

The following are four hypothetical scenarios indicating the kinds of activities that would be generated by teams within a PEN. Although these are hypothetical, they are drawn from real experiences of IUCN field projects. They are, in effect, composites of activities that can be found in a number of field projects.

Scenario 1. A PEC Project

Two anthropologists and a sociologist-demographer from the national university had been working informally with some of their students for some months with a small, isolated community on the edge of a newly defined protected area. The academics were included in the list of potential network participants and attended the PEN workshop. There they met the staff of the environment NGO who knew of the protected area, and who were assisting the government agency in charge of managing the area. In addition, the local FPA representative spoke of their work in that part of the country, and was interested in doing something jointly with the university team.

As a result of their meeting and discussions at the workshop, they decided to put together a proposal for a PEC project. The aim was to help the community organize to identify its problems and work out ways to address those problems. The team would be facilitators, and would especially help to establish links to the government and NGO community for resources the village might decide it needed.

The first six months of the project would involve working through a series of Participatory Rapid Appraisal (PRA) and planning activities with the community members and agreeing upon a plan for the activities they considered important. This would be followed by an additional 18 months to get the village plan in action.

The group put the proposal to the PEN office, which judged it to be of high quality and approved it for funding. PEN funds would go towards the time of the facilitators and preliminary field work. The facilitators would then seek additional funding to cover the activities themselves.

By the end of the first six months, the PRA had been completed, including a census and population projection using DEMPROJ, a (micro-computer) PC-based projection software. A village committee had been established and a number of initiatives were decided upon. The women planned a small water project which was to bring a single standpipe into the village. This would save them the walk to the stream for water. Since this pipe could bring water from a small dam farther up stream, the water would also be better quality than that near the village. A terracing project was begun on a nearby hillside that had been denuded and was beginning to erode. The environment NGO staff helped plan the water service and assisted in making a request to the regional government. The NGO staff also obtained the assistance of the agricultural department for the terracing project, and advice on a selection of crops suited to the location.

At this time, the boundaries of the protected area were not clearly marked out and the villagers were unsure what the protected area would mean for the game and forest products they had been gathering from the area

for their own use for generations. The environment NGO staff offered to negotiate with the regional parks officer to lay the groundwork for a possible collaborative management agreement with the government. The situation was not urgent, however, since government resources for the protected area were very short, and it did not seem like much would be done in the way of extending the protected area for another year. Moreover, this village was the most isolated of all in the area.

The population projection showed that numbers would double within a generation, with no new land available, causing the village committee to consider some kind of family planning scheme, for which the women of the village were in strong support. The closest primary health centre was half a day's walk from the village, and even at that station there were no family planning services.

The committee decided to send two women from the village to the regional capital for training from the local FPA, arranged by the FPA member of the project team. The two women, themselves mothers of six and seven children respectively, returned to the village and established an informal family planning clinic with a two-month supply of condoms and oral contraceptive pills. Shortly after their return, an FPA field worker visited the village together with a medical student from the university who was doing a programme in community medicine. Together they provided elementary instruction in first aid, sanitation, identification of serious illness and family planning. They proposed, with the village committee, to request regular visits of the mobile health van which was being established by a UNFPA project in the district capital.

Evaluation

In the first project proposal, the PEN participants proposed that the initial process of performing the PRA be the first point of evaluation. In addition, they proposed possible outcomes including increased food production and soil conservation, village water, and (quite uncertainly at the time) development of a local family planning project and increased contraceptive use. They also proposed a negotiation with the protected area managers with the objective of establishing a joint management agreement to give the village both responsibility for resource management and continued ability to take resources from the area.

Scenario 2. Planning for a System of Parks

A university geographer who had been working on a GIS of the country's natural resources was invited to join the PEN, and attended the workshop. The IUCN representatives put him in touch with the national parks office and they began to discuss how the overall park system was emerging. They also talked with university and government demographers about the concentrations of people around the park areas, the dynamics of those populations, and the range of social services – primary health care, primary education, and family planning services – that would affect the population dynamics. From this a team emerged that included the university geographer, a demographer, the head of the national parks system, a government economic demographer from the department of statistics, as well as members from the health and education ministries. They proposed to establish a GIS of the overall system of national parks and protected areas, and to include population dynamics in the system. After the project was approved and funding granted by the PEN office, the country UNFPA and UNEP representatives expressed an interest in the project and wished to be on its steering committee.

The first six months were spent in generating a digitized national map, made up of regional maps showing park boundaries and both population concentrations and locations of social service outlets (primary health care centres and primary schools) around the boundaries. In addition to making the digitized map, government sources were consulted for data on population, land cover and demographics, education, health and economics relating to the districts identified by the population concentrations around the parks. That data was then entered into the computerized GIS. This also resulted in an identification of data gaps, which led to the next stage involving sample data collection in the field.

At the end of six months, the regional and national digitized maps had been prepared and the available data entered into the GIS. The UNEP and the USAID representatives promised the team a series of satellite images that would cover the past 15–20 years. These were ordered, and the imagery depository was searched for an appropriate set of relatively cloud-free images to provide a time series of at least ten years. The data gaps had been identified and the team proposed another year of field work to collect data, especially on population, health and educational conditions in the remote areas. For the most remote areas, the team proposed a sampling process to provide ground truthing for the most recent satellite image that the team now had available. From this, they hoped to be able to devise a process for estimating population sizes and making projections from the satellite imagery. This information would then be used for the overall planning of the system of national parks.

The parks director was especially interested in identifying areas of high population concentration and rapid growth on park boundaries, and doing something about the pressures, so he formed a small task force consisting of himself and members of health, education and family planning services. This enabled them to plan suitable development efforts, and extension of needed social services in critical areas around the parks. The director was also interested in developing collaborative management agreements that would give local populations a stake in the parks and enlist their aid in managing them. For this he called on the assistance of the IUCN Social Policy Service, which was developing a series of collaborative management initiatives.

Evaluation

The initial objective was to provide the national parks office with the tools to plan for the overall system of national parks, especially in view of the people living around (and even inside) the parks and protected areas. The establishment of the GIS with park boundaries and population data, including population projections, was the first outcome that could be evaluated. The second point of evaluation was to be the capacity of the planning team to generate a plan for providing needed health and educational services for the areas surrounding the parks. A rather less certain outcome was to be the combined use of field sample censuses and satellite imagery to estimate population dynamics, especially in remote and sparsely settled rural areas.

The long term objective, in which the GIS was but the first step, was to assist the national parks office to plan for a system of parks that included local populations. Rather than attempting to exclude these local populations, the park's plan focused on joint management agreements, local development activities, and the provision of health and educational services that would give the local people some stake in the parks. It was recognized that evaluating that objective could not be done effectively in the

near term. Even with a well grounded GIS, it would take at least a year of discussions with the various government offices involved just to begin planning for parks and people together. The actual outcome of the plan – increased human wellbeing and natural resource protection – was more difficult to evaluate and could not be expected from the project for another few years.

Scenario 3. A Demographic Projection for the National Park

A university demographer from a remote northern region was invited to be a participant and attended the first workshop. He talked with the biologist from an IUCN-member NGO and learned of a problem they perceived in the national park located near his family home. The park had been in operation for many years, and was being managed by the national department of environmental protection with assistance from the NGO. The NGO was being called upon to help assess water quality and its impact on the bird life of the park's major lake, which was the temporary home of a large migratory bird flock. The NGO staff knew that the human population around the park had grown substantially and was beginning to press on the park itself. Poor farmers scratched out a meagre living on soil not suited to the crops they cultivated. A town bordering the park had grown rapidly and it was evident that pollutants from human wastes and chemicals were causing some damage to the wildlife. There was also an increased demand for wild meat. The DEP park manager wanted to know the precise nature of the population growth both in the vicinity of the park and in the town itself. He also wished to see a decline in population growth, since he thought it would ultimately overwhelm the park. His district, however, was strongly Roman Catholic, and the local religious authorities followed the recent strong leadership of the Pope against any form of artificial birth control.

The demographer, the DEP and the IUCN member proposed a small project and presented it to the PEN office for funding. The demographer took some of his students on a field trip and did a demographic analysis of the region, including the population around the park and in the major town. They prepared an estimate of the age–sex distribution and of infant mortality and total fertility rates. From this they then did a series of projections for the next fifty years. They also did a fifty-year projection with the base 20 years ago. With IUCN assistance, the assessment of population growth of the town would be linked to assessments of pollution of the lake and its impact on the bird population. The project thus provided field training in demographic analysis and projections for the university students. It also provided the DEP park managers with a vision of the future, which could be used both for park management and for planning for future resource needs. Finally, members of the IUCN Commission on Education and Communication would be able to use the information to add to a project on environmental education, which they had been developing with national and district education officials for the past year.

After six months, the project had developed the population projections and made some projections about lake pollution. The materials were being used in the schools and were generating considerable interest among the students and their parents. The next steps were somewhat uncertain, however, since the local family planning association activities had been roundly criticised by religious leaders, and there was little improvement in contraceptive use. Nonetheless two local women's groups were interested in learning more and were beginning a quiet campaign to promote some form of family planning. The district FPA worker suggested they might adopt the kind of natural family planning programme that had been successful in Mauritius, and this was actively being considered. Population policies at this local level had not changed, and it appeared that the more extreme of the demographic projections might be realized. However, with the work of the two women's groups, supported by the university demographer, it was likely that there would be a change.

Evaluation

Given the delicate nature of the local situation, the project was narrowly focused on obtaining demographic information, making projections, and making the projections with various scenarios accessible to teachers, environmentalists and women's groups for their environmental education programmes. On these grounds the project accomplished its aims fully, though the impact of these on the wellbeing of either people or the environment was not clear.

Scenario 4. National Population Environment Development Planning

Technical officers from the ministries of environment, agriculture and health were identified as potential PEN participants and attended the initial workshop. There they met a newly appointed economist at the university, together with one of his demographic colleagues, and were brought together in one session with representatives of UNEP, USAID and UNFPA. They were all intrigued with the potential power and utility of the IIASA Population–Development–Environment (PDE) model, and discussed ways to develop such a model for their country.

They decided to make a proposal to the PEN office in which the economist would bring together a small group of academic members under a national ad hoc PDE planning group to see if it would be possible, and what would be required, to adopt and adapt the IIASA model. The three officials from agriculture, environment and health undertook to provide leadership for the ad hoc group, and they invited representatives of relevant UN agencies and three bilateral aid agencies to attend the initial meeting. They also invited the national IUCN officials, and the director of the national Family Planning Association. A number of other national NGO leaders concerned with development, health and women's affairs were also included.

The PEN office provided a grant for a one year's planning effort, which essentially covered some of the university professors' time, as well as some support for meetings. An initial meeting was held to acquaint the various potential players with the idea. For this meeting, IIASA sent a representative to make a presentation of the IIASA model. UNFPA provided funding for that consultation.

There was sufficient interest in the model to take the next step of planning for its development in the country. To do this, the three government leaders worked with the academic team to plan a two-year project that would be located in the university, but would include government, NGO and international and bilateral aid donors. The larger group would meet monthly for the first six months to advise the academic team on data sources and the kinds of questions that should be addressed. For the next year it would meet quarterly to assess the progress and provide what assistance it could. The three leading officials approached UNFPA for funding to involve the IIASA staff to assist in developing and adopting the model.

By the end of the first year, the university team, together with an IIASA consultant, had laid out the specific modules of the model that would be utilized, and was busy collecting data for it. The university had created an interdisciplinary graduate seminar between four faculties, which now had 40 students working as research assistants on the project. The national level ad hoc planning group had been formally established as the PDE planning board, which was advisory to the prime minister and the central economic planning board. It would take another year of data gathering to put together all the pieces required to run the model and test different scenarios, but some interim partial scenarios were already near completion. Population growth with age and sex specifications could be projected for the next fifty years under a variety of assumptions. Each one was connected to the education segment, so that the implications of various patterns of population growth on education demand and costs could be roughly calculated. Energy consumption patterns had been estimated for all sources, including fuelwood. Thus population growth scenarios could show implications for different kinds of energy use, and forest depletion for fuelwood, alternate demands for fuelwood from plantations, and the foreign exchange implications of various energy use patterns – of imported fossil fuels for example – could all be partially calculated with some rough assumptions.

The PEN office continued to support some of the university's costs over the next year. UNFPA supported the IIASA consultation. Various bilateral donors supported different components, each of which required extensive data gathering: a forest component, a water component and an agricultural component. UNEP planned to provide a series of satellite images that would cover the country in six month cycles for the next few years, and would provide images of the past ten years as well.

Evaluation

The initial proposal aimed at exploring the possibilities of developing a national PDE modelling activity. It was to be evaluated by the interest generated in a number of donors for the project. More specifically, the proposal expected that both interest and human and financial resources would be forthcoming from a variety of national and international sources. For the long term, the PDE model would help provide a better system of national planning and mobilization and direction of resources to increase the wellbeing of both people and the environment. It was also recognized, however, that planning and government leadership were only one part, and often not even the most important part, of the determinants of such progress. Natural calamities and international market pressures could wreak havoc on the best of plans. Nonetheless, it was proposed that national planning, especially when bringing together a variety of specialist governmental, non-governmental and international groups, would certainly help to promote a kind of development that could be called sustainable.

Part III

Linking Population and the Environment – Frameworks and Models

To make substantive linkages between population and environment it is useful to have 'maps' or schemes that locate specific population and environmental conditions and show the pathways between them.

Frameworks are one type of map. They give us general categories of conditions (deforestation, poverty or women's status), and suggest some of the possible pathways that link them together, and to population and the environment.

Models are more elaborate maps, or more precise tools, that lead us to quantify specific conditions, and the strength and direction of linkages between those conditions. Models also permit us to specify conditions and run those out into the future, as scenarios, so we can see the possible outcomes of the conditions we now experience.

This chapter will review some commonly used frameworks and models for population–environment relationships. It examines a variety of uses for the common I=PAT framework, which has seen extensive use for the past two decades. It also provides details for one of the current models, IIASA's Population–Development–Environment model, which shows great capacities for providing practical guidelines for integrating population into strategies for sustainability.

Introduction

To understand how population conditions are linked with environmental conditions, it is useful to have a set of maps that identify specific population and environmental conditions and show the connections between those conditions. These maps can help us to answer such questions as the following.

- How is the size or growth of a population linked to the quality of water, the extent of forests, or the condition of the soil?
- By what route, or through what kinds of activities, does human population growth lead in some cases to the reduction in forests and in other cases to their increase?
- Under what conditions do the wastes generated by human living lead to sickness and death, or to health and wellbeing?
- What impact does the age and sex composition of a population have on its levels of production or consumption?
- What impact have laws and policies in health and education had on population, production and consumption, and the environment?

With a good set of maps, and some knowledge of what we have done in the past, we can answer these questions, thus increasing our understanding of our present condition. But we want to do more than simply understand. We want to make things better. A Strategy for National Sustainable Development implies that we will actively intervene to make the future better. Many countries and groups are attempting to plan directly and rationally for sustainable development. To plan for the future, we also need to ask somewhat different questions.

- What will the future look like if we continue to *produce*, *reproduce* and *consume* as we do now? And what would the future look like if we changed our current patterns of production, reproduction or consumption?
- What are the future implications for human health of recent rises in the use of fossil fuels? What will be the future impact on health if we increase or decrease fossil fuel use?
- What are the future implications of continued fertility levels, or of rising or falling levels of fertility? What are the implications for health, for education, for production and for environmental protection?
- What are the future implications of current rates of deforestation, or wetland development? What impact will these trends have on human health and wellbeing?

To answer these questions, we need the capacity to use the many complex connections between population and environment in order to envisage scenarios for the future. For this we need models that enable us to extrapolate the population–environment linkages we can now specify. We need models which will allow simulations of the future.

As we noted in Part I, none of the linkages between population and the environment are simple and direct. All conditions of population work through human forms of social organization and technology to have an impact on the environment. Similarly, all conditions of the environment work through some form of human–social organization or technology to have an impact on the size or conditions of a population. The link between human wastes and health in a country, for example, will depend on the numbers of people, the type of wastes produced, the infrastructure for water procurement and waste disposal, and the infrastructure for health maintenance. It will also depend on the level of nutrition, and thus on the kind of food production a country has, as well as on world food prices. If this were not sufficient, one must also add wastes that intrude from other societies. The ozone depletion that increases the incidence of eye disease and skin cancer in the southern hemisphere comes largely from past emissions of chlorofluorocarbons from the northern hemisphere. Thus even the relatively limited health–waste linkage is a highly complex one involving many natural resources, human technology and forms of social organization.

Characteristics of Frameworks and Models

We now have at our disposal a set of tools that give us the capacity to trace out many of the population–environment linkages. Here we will make a somewhat arbitrary, but still useful, distinction between frameworks and models. By frameworks, we refer to complex sets of ideas suggesting how different conditions are linked together. We shall make a further distinction between two commonly used types of frameworks: sectoral and institutional. In the first, we look at different sectors of the economy, such as agriculture, forestry, mining, industry, energy or transportation. For each sector, certain conditions can be described, such as the need for labour, or the provision of goods, implying arrows that connect one sector with another. In the second, we identify specific human institutions that influence the relationship between population and the environment. Here one can trace connections between markets, political systems, women's status, values, and the international economic order on the one hand, and certain outcomes for the environment on the other. We shall have an opportunity to see both of these types of frameworks.

Frameworks

Frameworks are useful in that they suggest how specific conditions might be connected with one another, and how long, indirect trains of connections might lie behind any specific environmental impact. Their use lies in suggesting connections that we might otherwise miss, or raising questions about how

the current conditions of human life or the environment might affect one another. They are suggestive and illustrative and thus can lead us to ask searching questions about how population and the environment are linked in an overall national setting.

Frameworks, as we are using the term, are not, however, highly precise statements of either conditions or their connections. Thus they cannot provide specific answers to questions of what might happen in the future under given conditions. For those questions, we need greater specification of both conditions and linkages and for those specifications, we need models.

Models

Models have been developed that state the precise value or character of any human and environmental condition, and the direction and strength of the connections between those conditions. Today the computer has greatly facilitated the process since it can perform massive and complex sets of mathematical calculations with great speed, making it possible to run future scenarios with ease and with useful sets of numbers and graphs. Computer-driven future modelling has greatly increased our capacity to see future implications of current conditions, and also to examine possible future implications of deliberate interventions we can make now.

The most elaborate models now in use are the *Global Circulation Models* used to chart the links between greenhouse gas emissions and climate change. These employ thousands of equations with data taken from thousands of points of observation on land, in the sea and in the atmosphere. They require massive computer capacities and can chart possible future scenarios in temperature and precipitation for the world as a whole, and also for very large regions. They have been used by such groups as the Intergovernmental Panel on Climate Change (IPCC) to assist in planning international conventions that might mitigate the human impact on the environment. The Global Circulation Models provide the most advanced and sophisticated of the population–environment models, but they are not very useful for national levels of planning. They are as yet too crude to be able to predict climate changes for other than very large areas of the planet. For national level planning, we need somewhat simpler models, which work on a smaller scale.

Models and modelling are not without their shortcomings, and their critics, however. And some of the criticisms are quite severe. Two are of special importance for our purposes:

1. Simplification

One major criticism focuses on both the strength and the weakness of any model: its simplification of what are always highly complex situations. No model can, and no model attempts to, capture all the complex connections, conditions and movements in any real life situation. Models work precisely because they simplify reality. It would not be possible to build a model that included all the conditions of any action. Even the simplest human behaviour, for example, contains conditions that move from subtle chemical reactions and neuroelectrical impulses to muscular action, past psychological conditions,

current numbers, actions and conditions of immediate other humans, micro-economic conditions, global economic conditions, local weather, land use patterns and global climate conditions. And that is only a highly abbreviated list. The number of conditions that could be quantified and built into any model of human behaviour is almost limitless. There is no way that all of this could be built into a single model. But modellers argue that there is no need for all conditions to be included. A selection must be made of just a few conditions; reality must be simplified if we are to comprehend it at all, and especially if we are to attempt to use models to help us look into the future. Such simplification is, after all, very common in most of what we do. We use a simple measure of body temperature to provide a prediction of how well we will perform. We use simple measures of body weight to tell us how well children are performing. We can predict quite accurately how millions of people will vote in an election from a simple random sample survey of just over 1,000 people. We can predict how people will respond to changes in prices from knowing just a few things about prices and consumption. We can predict the size and composition of a future population by knowing a few things about the current population. We can make these predictions not because we know everything, but because we have selected a few critical conditions and have generated good information about them. Simplification is, by necessity, common in the way we think and act.

In modelling, simplification allows us to quantify a limited number of conditions and their dynamic interconnections. With this we can ask a series of 'what if' questions. Answers can be derived from the model for periods into the future. Of course, the answers are only statements of probability – if we change this policy, or make this investment, this is what is likely to happen, if we have included the right conditions and have the right quantities for both conditions and connections. That caveat leads to the second major criticism of models.

2. *Getting it right*

The utility of any model depends on its accuracy (or validity, to use the more technical term). This applies both to current conditions, and to the relationships between conditions. Does a population model include the right numbers of people, the right distribution of males and females and people of different ages? Are the birth and death rates by age, sex, education, and income correct? Are assumptions of future changes in birth and death rates, especially with expected changes in economic conditions, correct? If the quantities of current conditions are wrong, it is not likely that future predictions will be right. If we do not have the correct relations between current conditions, such as death rates and income, we will not get valid predictions.

One of the serious disadvantages of a quantitative model lies in the fact that its output will always be given in very precise figures, even if they are wrong. This was noted above in respect to population projections, but the same is true of any modelling. Because the predictions are stated in numbers (technically called interval variables), they can be stated in a very precise form, to many decimal places. Their precision gives them an air of accuracy which they may not deserve.

One way to deal with the problem of accuracy or validity is to construct a model covering past conditions. For example, the WORLD3 model used in *The Limits to Growth* (Meadows, 1972) and *Beyond the Limits* (Meadows, 1992) models the period from 1900 to 2100. When all quantities (of conditions and relations between conditions) are entered, the model is run for the period up to the present, 1900 to 1990. The output can then be compared with what actually happened to validate the model. In this case, the experience of the past 90 years can be used to determine whether the quantities in the model are correct, and adjustments can be made where necessary. Once the model is validated, we can have more confidence in the validity or accuracy of the predictions.

The confidence will never be complete, of course. There may still be errors in the quantities, and all manner of unanticipated changes can appear in the real world, which will make the model's predictions quite wrong. This is not a reason to reject models; however, it is an argument to use models with intelligence and sensitivity.

Here we shall examine one model that shows special usefulness for SNSD; the IIASA PDE model developed for Mauritius with funding from the UNFPA.[3]

Let us begin with the simplest framework, then move to progressively more complex models.

3 UNFPA has been especially innovative in supporting the development of computer simulation models, which permit the generation of scenarios from chosen current conditions. It supported the development of the FAO Computerized System for Agricultural and Population Planning and Assistance and Training (CAPPA), and the IIASA model.

Common Frameworks: I=PAT

4 See also Ehrlich
and Ehrlich, 1990
and 1991.

5 This is called an
identity because it is
true by definition. In
the equation the P
and C values cancel
each other, and we
are left with the
statement: I=I.

I=PAT

One of the most well used frameworks is the I=PAT equation, initially suggested two decades ago by Paul Ehrlich and John Holdren (1971).[4] The equation simply says that the impact on the environment is equal to Population times Affluence times Consumption. Another way to write this is an identity, as follows:[5]

$$I = \frac{P}{1} \times \frac{C}{P} \times \frac{I}{C}$$

In this equation,
I is the impact on the environment;
P is the size of the population;
Affluence is indicated by Consumption per Capita, or C/P; and Technology is indicated by the level of Impact per Consumption, or I/C.

The framework at this level asserts that population growth cannot continue without having a serious impact on the planet, because consumption per capita and the impact of technology cannot be infinitely reduced. As Ehrlich stated recently in rather graphic form (1994),

...sustainable development cannot continue without limits to population growth....Even raising human beings as chickens are mass-produced would require some minimum flow of energy and materials per person (A), and the second law of thermodynamics dictates absolute limits to the efficiency with which technology (T) can maintain that flow.

Although Ehrlich's main argument has focused on the necessity to limit population growth, the identity can also be used to argue for limiting consumption or the impact of technology. It can also be turned to show that poor countries, with more rapid population growth, have a less damaging impact on the environment than do rich countries with high levels of consumption and more destructive technologies. Thus one of the main values of the I=PAT framework is to show that population is not the only condition that produces environmental impact; production and consumption are also important elements. This recalls a statement made above, taken from IUCN's *Caring for the Earth* (1991):

The Earth has limits... To live within those limits, two things need to be done: population growth must stop everywhere and the rich must stabilize ... their consumption of resources. (p 5)

In this form, the framework is of limited value, however. It merely asserts that population growth, consumption, and technological impact cannot continue to increase without reaching the limits of the carrying capacity of the earth. It has been used in other ways, however, which have increased its capacity to show what the connections are between population and the environment.

Clark's adaptation: National Historical Experiences

In 1992 William Clark published the results of a twelve-country historical analysis based on the Ehrlich framework (1992). He focused on the production of carbon dioxide as the impact on the environment. This is a useful surrogate for environmental damage, since the rise of carbon dioxide emissions comes directly from the emergence of the new urban industrial society of the past two centuries, and this rise is generally believed to portend long term warming of the planet.

Clark asks how much carbon dioxide emission has been produced over the past 50 years in each of 12 countries. He also asks what are the relative weights of the three components – population, affluence or consumption, and technology – in producing this level of carbon dioxide emission. Clark chooses 12 countries to represent different types of economic systems or levels of development. Canada, Japan, the UK and the US are used to represent the highly industrialized countries; China, Poland and the USSR represent the centrally planned economies; Brazil, India and Indonesia the poor developing countries; and Kenya and Zaire the more stagnant poor countries. The figures Clark uses for each of the elements in the equation are readily understandable, and generally available.

For population, he uses the simple growing numbers, with no distinction by natural increase or migration, and no specification of age or sex distributions.

For affluence or consumption, Clark uses the simple aggregate figure of GDP per capita. This also equates consumption with economic development, making it relatively easy to measure given current data systems. It also has the advantage of using a figure that most nations use to measure their success in a major national goal, achieving economic development.

For technology Clark uses an estimate of the amount of carbon dioxide produced per unit of GDP. On the national scale, this can be thought of as the energy efficiency of

technology, and is currently an important measure for thinking about energy and sustainability (Smith, 1993). The less efficient technology we have, the more energy will be required and the more carbon dioxide we will emit per unit of wealth. Historically, it has been found that countries are less energy efficient in the early stage of development, but become more energy efficient as they become more developed (see the following Box). Thus this measure of carbon emissions per unit of wealth is a key link in the relationship between population, environment and development.

The results of Clark's analysis should come as no surprise. No single condition – population, consumption or technology – was the major cause of environmental impact in all cases. Each dominates at some time and place for the countries examined. For example, since 1955, population growth has been the dominant cause of increased carbon dioxide emissions in Kenya and Zaire. Increased consumption, or economic development, has been the major force in Japan and China. And increased energy efficiency has been the dominant condition, reducing carbon emission growth rates, in Canada and the US.

There are two important lessons to be learned from this exercise. One reinforces the basic point made earlier; population–environment relationships are location specific. There are few hard generalizations that can be used everywhere. Each country will have to determine which of the components is currently having the largest impact on the environment, and from that can identify appropriate interventions, policies and projects.

The second lesson is that this is a relatively simple exercise that any country can undertake, examining both its past and its near term future. Population and GDP data are generally available for the past and are usually projected for the next five to ten years in

A Key Linkage: Energy, Population, Development and Environment

Energy use appears to threaten sustainability.

It was a revolution in energy – the rise of fossil fuels – that brought us the urban industrial society that promises a better life but also poses major challenges for sustainability. To achieve the economic development that all countries now desire and to provide for the ever increasing numbers of people on the planet requires more energy. But energy consumption produces the kind of environmental degradation that threatens sustainability. It increases pollution and the emission of greenhouse gases, threatening major climate change. Current levels and rates of growth in energy consumption are not sustainable.

Energy efficiency holds a promise for sustainable development.

It is true historically that as countries begin economic development, they seem to need more energy to produce each new increment of wealth or output. Energy consumption grows more rapidly than wealth production in the early stages of development. But it is also true that in the process of development a peak is reached in levels of energy consumption per unit of wealth; and from that point technology becomes more efficient, and the amount of energy required to produce an extra unit of wealth declines. The UK reached that peak around 1860 and the US around 1920. All industrial countries today show a decline in the amount of energy needed to produce a unit of additional wealth. Moreover, the peak of energy inefficiency appears to be declining. It was lower for Germany than for the UK and the US, and lower still for France, Japan and Italy.

Increasing energy efficiency is a key requirement for poor and rich countries to promote the wellbeing of both people and ecosystems.

Source: Smith, 1993

national development plans. There are also estimates of total carbon emission (from both fuel consumption and land cover changes), from which annual interpolations can be made to provide a rough estimate of energy efficiency. With these data, countries can develop future scenarios, showing what they could expect from population growth, economic development, and the changing technology that affects energy efficiency.

Harrison's Institutional Framework

The I=PxAxT identity has proven to be a useful framework for illustrating the linkage between population, development and environment for both individual countries and for the planet as a whole. It does have limitations, however. One is that population is considered as only one condition – the growth rate of the total population. Similarly, both affluence and technology are expressed as simple ratios indicating one aggregate condition, such as economic development or aggregate energy efficiency. They say nothing about the human institutions that lie behind the ratios. It is well known that GDP per capita says little or nothing about the distribution of wealth, and the simple ratio of carbon emissions per dollar of wealth tells us little about the kind of institutions and technology we are working with. It is possible, however, to develop larger and more complex frameworks to identify important conditions of human social organization that lie behind all three elements of the basic identity. One of these has been developed by Paul Harrison in his important book, *The Third Revolution* (1992).

Harrison's broad framework is shown in Figure 3 (opposite). It posits a system in which a series of institutional arrangements are interconnected and there are no ultimate causes. Population, consumption and technology are treated as one set of interconnected conditions that act on the environment by extracting resources and emitting wastes, causing environmental depletion, pollution or degradation. This produces signals that work through a variety of social conditions to affect

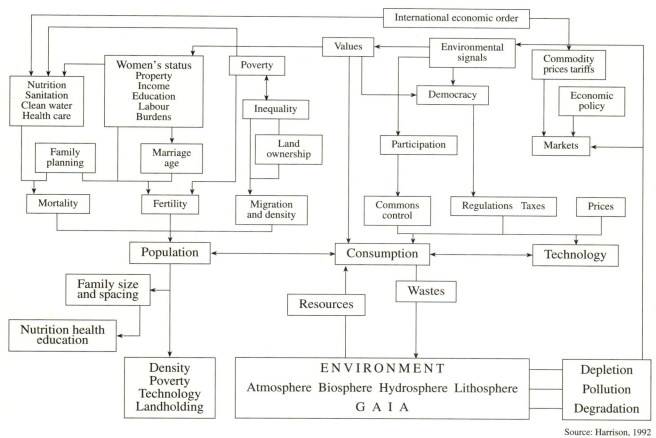

Figure 3

Source: Harrison, 1992

population, consumption and technology to complete the circle.

One can trace through the boxes and arrows of Harrison's chart to see the specific connection he believes are important, and how those connections operate. The human use of the environment (top right hand corner) produces signals that affect policies (commons control, regulations and taxes) and price changes that in turn affect consumption and technology. The set of paths moving downward on the right hand side suggests, for example, that democracy and participation speed up the impact of the signals on consumption and technology to reduce the degradation. Values also play a role in determining how much democracy there will be. Note, too, that the international economic order also impinges on a national system. This is an external force that all governments will have to deal with. On the left hand side of Harrison's map, the environmental signals flow through values to affect women's status, which in turn affects a series of other conditions that ultimately affect the size and growth rate of the population. Harrison also suggests in this framework that the international economic order produces poverty, which has an impact on population through such things as inequality, migration and fertility.

These are all important observations, and in any national context could lend themselves to useful questions. What is the status of women in a given society? How does it affect fertility and population growth? How does it affect production and consumption? Further, what is it that determines the status of women; is it values, or some other economic condition? Asking these questions can also lead to a consideration of policy changes that might affect the status of women and therefore lead to a better population–environment relationship.

It should be emphasized, however, that these frameworks should be used to raise questions about a specific national setting. The process of raising questions is most important, since the answers to any of these questions will not be universal. They will all be location-specific, and in themselves will show how other conditions affect any given outcome. For example, the international economic order, working through trade and aid, has been associated with increasing poverty in some countries (e.g. in sub-Saharan Africa), and with increasing affluence in others (e.g. Japan, South Korea, Taiwan). Understanding how the international economic order operates implies knowing why it produces poverty in one location and affluence in another. What national policies made some countries rich and others poor? Obviously there are differences in national policies, or responses to the international economic order, which operate at the national level and are thus the responsibility of national governments.

The same can be said for the increasingly critical issue of women's status and roles. It is clear that institutional conditions that discriminate against women and girls reduce the capacity of a society to promote the wellbeing of both people and ecosystems. But the ways that societies reduce discriminatory practices and enhance the position of women must take into account specific national and even subnational conditions. Broad goals can be stated in such things as the 1966 Teheran United Nations Declaration of Human Rights, but the way those will be translated into specific actions and policies must always be specific to a national government.

It should be noted that in Harrison's map (or any other similar framework) these are linkages that he *proposes* are operating. They are in effect hypotheses. Harrison provides some evidence for them in specific locations, and many will be familiar and accepted more or less as the received wisdom. For example, in communist countries the lack of market freedom, competition, private property and democratic rights is stated as the reason that '...environmental degradation has been particularly bad.' This is a position that Harrison illustrates with specific examples, but it is not what would be called a well proven rule. That is, we can also find environmental degradation in open, free market situations with extensive private property. Many institutional conditions can slow down, speed up or in other ways distort the signals that emerge from the population–environment interaction.[6]

In addition, the diagram itself does not always clearly identify the condition and its connection to other conditions. The 'markets' box, for example can mean many different things. In the narrative quoted above, it means a free market, and markets can be defined by the extent of freedom or imperfection. But there are different dimensions of market freedom, and there are other ways to describe markets as well, which would include other boxes, such as economic policy and tariffs. Nor does the diagram tell us whether a particular

6 For a good discussion of the problems of signals in a system, see Meadows, Meadows and Randers *Beyond the Limits*, 1992.

pathway, or arrow, is having a positive or negative impact, though often Harrison provides these suggestions in his narrative. We have already noted that there is a path from the international economic order to poverty. Is it positive or negative? One can easily find examples of both. International trade, aid and investment were certainly forces in increasing the wealth of Japan, South Korea, Taiwan, Singapore and Thailand. But one can make the case that structural adjustment is currently digging Africa into deeper poverty. The arrow is suggestive, however, and would lead one to ask what is the impact of the international economic order in any national case.

These are common shortcomings with frameworks as we have defined them. There are others that we review below. Nonetheless, they can be useful ways to illustrate important connections. And Harrison's diagram offers an important example of the kind of map that can be drawn to show how population, consumption and technology are affected by different institutional arrangements and through these have an impact on the environment.

FAO's Sectoral Framework

FAO is undertaking a series of papers on population and the environment, aimed at bringing state of the art knowledge to their field operatives and to the UNFPA Country Support Team Advisers. The first of these papers (Marcoux, 1994) focuses on water resources, easily considered the most important of all natural resources. The paper reviews population dynamics affecting the demand for and the supply of water, and the impact of water supply on population. Marcoux then uses the I=PAT framework to examine population–water linkages.

Here the framework focuses on three sectors where the population–water linkages

show different or distinct kinds of problems: domestic water use, agriculture and industry.

Domestic Water Use

Here the Impact is equal to Population size times (A) consumption per capita, times (T) withdrawal. For simple future scenarios, one can assume different levels of population growth, consumption, and withdrawal. Even more useful, however, is to ask what affects levels of water consumption and withdrawal. Per capita water consumption varies considerably, and can be compared with the WHO standard requirement of $11m^3$ per year (30 litres per day) for domestic use. The world standard is currently about 52 m^3, but regions range from a low of 17 in Africa to a high of 167 in North America. Individual countries show an even greater range. For national level planning, one can consider such things as family income (wealthy to poor) and location (rural/urban or other regions) as major determinants. Technology or withdrawals can refer to the public utilities infrastructure that gives individuals access to water. By first examining local conditions, such as how family income affects water consumption, or how public infrastructure investment affects withdrawals and consumption, a future scenarios exercise can be generated to show the relative impact of population growth, changing consumption and changing technology on the environmental impact on water. Such an exercise can also indicate levels of investment required to meet given consumption levels under projected rates of population growth.

Agriculture

For agriculture the (A) affluence element is indicated by food consumption per capita, while (T) indicates the demand for irrigation for the food. Future scenarios can be developed with different assumptions of food consumption, and the different demands for irrigation for different crops. This type of scenario development can be further specified for each crop, especially for crops that have different irrigation requirements. Finally, different scenarios can be developed for different assumptions of domestic production versus import of crops. The crop specification of the scenario offers many opportunities for examining the relative impact of population and significant changes in consumption and technology, all of which are subject to manipulation through national planning.

Industry

For industry, (A) is the consumption per capita of a specific product, and (T) is the water requirement for that product. As in agriculture, the product can be further specified as that domestically produced versus that imported. It is also well known that the water requirements of specific products vary greatly. For example, computer chip production requires immense quantities of high quality water; financial institutions in moderate climates require very little. As Marcoux notes, there can be an I=PAT calculation for each product. The product mix of industrial consumption can vary greatly and can become an element of national planning. Thus the I=PAT framework can be used to calculate the relative water impact of population, affluence and technology in a wide variety of sectors with a wide variety of specific conditions and products as the determinants of the impact.

Limitations of I=PAT

In all the cases treated above the I=PAT framework has uses, but also important limita-

tions. One of these concerns the definition of P or population. Lutz and his colleagues at IIASA have recently given a dramatic example of this (MacKellar et al, 1995). Typically the P in the framework stands simply for the number of persons in a population. But households are also significant units of consumption; the number of households is growing in the more developed regions, but remaining constant in the less developed regions. This comes in part from ageing and the increasing tendency for the aged in the more developed regions to live alone. But is also comes from divorce and reduced marriage rates. The point here is that if *persons* is used as the population measure, it counts for only about a third of expected CO_2 emissions over the past 20 years, but if *households* is used, population contributes three-fourths of past emission. The authors introduce an important caveat:

> *Should the unit of account be the individual, the household, the community, or some other component? Until more is known about the nature of the activities which give rise to environmental impacts, the answer will not be clear.*

I=PAT is a useful first approximation, and calls attention to the number of conditions that produce an environmental impact. But it is just a first approximation, and probably should not be used for serious policy formulation.

A more serious limitation is that it essentially solves the problem of the linkages by definition. Since the framework is an identity, by definition, population, affluence and technology cannot all increase together, without increasing I, nor can any one continue to increase, since the others cannot be reduced to zero and below. The framework is thus useful to call our attention to the limits to growth, but it cannot say where those limits are or what makes them higher or lower.

Moreover, each of the elements – P, A, and T – is assumed to be independent. Thus we cannot ask the question of what impact population growth, for example, might have on either consumption or on technology. This is, of course, one of the basic controversies in population theory. Malthus assumed that population growth itself would reduce affluence. The major alternative theory in population–environment dynamics, that of Esther Boserup (or earlier of William Godwin), holds that population growth puts pressure on natural resources, and leads people to develop new technologies to generate a higher quality of life, either by opening access to new resources (e.g. new boats to fish out of the rivers in onto the seas, or shifting from dry to wet rice cultivation) or drawing more sustenance from the same resource (e.g. raising crop yields or developing preservative techniques to reduce waste). Whether such changes are possible, and what might promote or slow them, are questions that cannot be asked with the I=PAT framework. For that type of question, we need more sophisticated tools, such as those found in modelling.

Models of Population–Environment Dynamics

Models permit us to simulate conditions and thus provide us with the possibility of thinking ahead. We can look at the possible outcomes of current trends, and we can try to estimate what will be the outcome of either deliberate policy changes, or other conditions over which we might not have control. We can also look back at the past and estimate how conditions would have been different under different policies or interventions.

One of the especially useful functions of simulation models is to help avoid 'overshoot'. This is where the movement of a particular condition, such as population growth or water use which might be growing under favourable conditions, develops sufficient momentum that it is difficult to stop even when resources can no longer sustain the growth. Small amounts of pollution flushing into a lagoon may build up slowly until a critical point is reached, tipping a balance that kills the lagoon. Over-fishing a resource may not be noticed until a point is reached where a species is sufficiently depleted that it cannot survive. If we can see the future implications of the growth of a particular activity, we can avoid overshooting the resource constraints by slowing down the growth before we reach the limits.

Models can be very simple, or highly complex, but they all contain certain basic elements. They must have a series of initial conditions, like the size of a population, the wealth of a society, the extent of a forest, the amount of pollution, or the output of some product like corn or steel. They must also have a set of parameters, estimates or measures of

the quantitative relationship between the initial conditions. These are all set out in a series of equations. For example, one equation will show that a future population will be the result of the current population plus the birth rate, minus the death rate over a specified period. Then another equation will specify the relationship between such things as the wealth of the society and both birth and death rates, usually with parameters as a decimal (e.g. a one step rise in wealth or per capita output will produce a 0.08 decline in the birth rate). These parameters must be known from empirical studies, or they may be assumed or guessed at. Guessing makes simulations quite risky, since the guesses will often be no more than the preferences of the modeller.

Simulation models are designed to trace out future changes. But they usually begin with a period from the past. This provides data on real changes in the past with which to validate the model. The WORLD3 (Meadows, 1972) model, for example, begins at 1900, and runs through 2100. This provides 90 years of real experience to assess the accuracy of the parameters. The IIASA Mauritius PDE model that we present below had data from 1962, and ran to 2050. The model was thus validated with almost 30 years of actual data. Then, by running the model with different assumptions from 1962 to 1990, the team could suggest what might have been different without the population change or the economic policies that actually occurred in this period.

The complexity of the model is reflected in part by the number of equations it contains.

The well known WORLD3 model developed by Meadows, Meadows and Randers (1972 and 1992) is considered a medium sized model and contains 149 equations linking 225 variables in five sectors: population, agriculture, economy, persistent pollution, and non-renewable resources. It is a complex systems model for the world as a whole, with both negative and positive feedback loops in and between each of the sectors. Note that it is, however, a global model, and thus does not specify regional conditions or differences.

The largest and most complex simulation models currently in existence are the global circulation models, which attempt to predict future climate changes under different atmospheric and oceanic conditions. These contain literally thousands of equations, with parameters for relationships between atmospheric gases, temperature, and circulation for different layers upward in the atmosphere and downward into the oceans. They require the largest and fastest of the Cray computers and calculations can take days to complete.

Neither the WORLD3 model nor the global circulation models will be useful for national level planning for sustainable development. Returning to a position expressed frequently, population–environment dynamics are location specific. The impact of population growth or decline will not be the same everywhere. It will depend on a large array of different local conditions. National planning must be done with national models.

For a variety of reasons the burst of modelling for sustainable development that began with extensive discussion, and just as extensive controversy, in the early 1970s died out and made little progress for almost the next two decades. It is now being revived, however, in specific locations.

In the 1980s, the FAO and UNESCO undertook a series of studies, with UNFPA

Modelling a Sahelian Ecological Collapse

In the western region of Niger, on the border with Mali, a Sahelian nomadic livestock raising ecosystem has developed over centuries. Existence is precarious. Rain is often scarce and unevenly distributed. Nomads move herds in search of good grazing in a large area that is essentially a commons, with no private property rights. In periods of good rains, people and livestock build up, often overgrazing, which degrades the land, making the situation ripe for a collapse. In the early 1970s, the region suffered a severe drought. Thousands of people died, more became ecological refugees and the animal loss was substantial. This was a classic case of the tragedy of the commons.

In 1974 Anthony Picardi (1974) undertook a study to treat this 'tragedy of the commons' with an explicitly interacting ecological–social–economic model. This was done at the Massachusetts Institute of Technology, where other social scientists were developing the kind of systems models that the Meadows were using for their Club of Rome study, *The Limits to Growth* (1972).

Picardi's model contained an ecological segment on the Sahel nomadic system, a typical economic segment, a segment on natural capital, and an innovative segment on policies (taxation, food relief, health and nutrition programmes, well digging, and veterinary assistance). He obtained data for the model from as early as 1920, thus providing half a century of solid data with which to validate the model. Some of the findings are both chilling and compelling.

The root cause of the ecological and economic collapse of the 1970s was not the severity of the drought. There had been equally severe droughts in the past on about a 30 year cycle, without the great losses suffered in the 1970s. Simulations showed that the drought affected the timing of the crash, but not its severity. That was caused by the cumulative effects of a few years of good rains that brought a buildup of people and cattle.

Simulations also assessed the impact of interventions such as well-digging, public health and veterinary care programmes. These interventions worsened the effects of the drought, because they increased the speed of population and herd buildups. The no-interventions scenarios produced less loss of life and herds because the slower buildup of people and cattle led to less overshooting and less land degradation.

Source: Sanderson, 1994

funding, aimed at assessing carrying capacity and developing tools for planning sustainable development. In 1986, FAO published the

Overshoot I: The Condition

A farmer has a small pond for fish and for irrigating a small vegetable garden. He plants some water lilies to attract frogs and provide some nutrients in the water. Although it is not immediately apparent, the water lilies find the environment much to their liking and they double in size every day. One day the farmer finds one quarter of the pond is filled with water lilies and thinks that is enough. Tomorrow he will begin to cut them back. He didn't get back the next day, but the day after. That day, however, he found the entire pond covered with water lilies, and it was almost impossible to cut them out. The fish are crowded out and it is difficult to draw water to irrigate the garden. He overshot his goal and seriously reduced his resource base. Overshoot easily occurs when there is rapid and exponential growth.

Overshoot II: The Determinants

A driver in a car sees a stop light ahead, moves his foot from the accelerator to the brake, and brings the car to a stop at the light. He did not overshoot. It was avoided by:

a clear signal:	the stop light,
perceived:	the driver saw the light,
acted upon:	moving the foot from the accelerator to the brake
in time:	foot movement and brake were fast enough for the speed of the car
effectively:	the brakes on the car worked.

If any of these five conditions had not been in place, the driver might have overshot the stop. If the signal had not worked; if the driver had not seen it; if his reaction time had been impaired (e.g. by alcohol); if the car had been going too fast; if the road had been icy; or if the brakes had not worked, there could have been an overshoot. If the stop light had been at a busy intersection, with many heavy trucks and cars going by, the results of the overshoot might have been disastrous.

results of an assessment of the capacity of world regions to feed their population in the next century (FAO 1986). This study, carried out with IIASA, examined soil quality, slope characteristics, degradation vulnerability and climate and estimated food output under low, medium and high inputs. Although the study did include estimates of population growth for the regions, the treatment of inputs was not linked to other parameters in feedback processes. This left important questions regarding the source and cost of inputs unaddressed. Following this, UNESCO and FAO collaborated in the development of a new, more dynamic systems model, based on energy requirements. The model was called ECCO – Enhancement of Carrying Capacity Options.

From the outset, the ECCO exercise chose to use energy units as the basic measures for the model. Most economic modelling uses prices or values for inputs and outputs. Although this makes the results accessible, especially in terms of economic planning, it poses a major problem in forecasting. Prices simply will not hold still, and predicting prices and exchange rates in the future is hazardous at best. The ECCO modellers thus decided to use energy units (King 1991). Resources are assessed in terms of the quantity of energy required to turn those resources into goods or services. This solves the problem of the instability of the units of analysis, but it introduces another. The results are not readily accessible. It is easy for people to understand the costs of resources, goods or services; it is not easy to grasp the energy units required. A validation of ECCO was run on the British economy from 1974–84 (BRITECCO). It performed reasonably well, with errors in sectors running from –20 percent in industrial exports, to near zero in agriculture, to +9 percent in investment-good outputs (Slesser 1987, 1991). Exercises were also developed for Kenya and for Mauritius.

While the ECCO approach has much to recommend it, especially in using firm units of analysis – energy requirements – it has a major disadvantage in its inaccessibility. Moreover, its focus is on carrying capacity which is, at best, a difficult concept to operationalize in any general form. Like sustainability, the idea appears simple in the abstract. Carrying capacity is the number of a species a particular habitat can support in relative stability. Sustainable development or sustainability implies the standard of living or resource use that does not bring down capital or resources for the next generation. Operationalizing these terms is difficult, however, because both depend, as we have said repeatedly above, on the social organization and technology that humans bring to the task of survival.

Thus a more useful modelling strategy would be one that remains very close to actual human conditions of production and consumption in a specific area. It is also apparent that modelling will be more useful if its parameters and values are those readily understandable by those who use the model for planning. These conditions have been well satisfied in the Mauritius IIASA activity.

IIASA's Population–Development–Environment Model[7]

7 We shall take some time and space to explore this model for a number of reasons. First is that we find it the most useful and powerful of the models now being used for population environment analyses and planning. In addition, although the model and its full story have been published by Springer-Verlag in Germany in 1994, the price of over US$60 will make it unavailable to many interested readers.

8 This is taken from the 7th century BC Greek philosopher, Anaximander, who is also credited with drawing the first map of the world. He proposed that the world is made up of earth, water, fire and air in some kind of balance (Lutz, 1994, p 213).

One especially effective model has recently been developed by IIASA in Laxenburg, Austria, in collaboration with the University and Government of Mauritius.

As important as the model itself was the process by which it was developed, which should be considered an essential element in this type of exercise. In the early 1980s the University of Mauritius developed a 'Mauritius 2000' study project. The approach was designed to be interdisciplinary, with biotechnology and population given prominent places. At the same time, the multi-disciplinary IIASA was examining population–environment interactions and proposing to revise the aborted modelling experiments that had started in the early 1970s. At the time IIASA had participated in the FAO Carrying Capacity study, and observed the ECCO model developments, with the incomplete experimental works done in Mauritius. This led Dr. Wolfgang Lutz of IIASA to propose a collaborative study which produced the Population–Development–Environment model for Mauritius in a project funded by the UNFPA. A team of 30 scientists was mobilized, including 7 from Mauritius, 11 from IIASA and 12 other international experts. Dr. Nathan Keyfitz of the US and IIASA, and Professor Jagdish Manrakhan, vice chancellor of the University of Mauritius, were the scientific supervisors. The UM–IIASA team that developed the model thus also built a substantial local capacity in Mauritius itself. This capacity continues to grow and the model is now in active use for future planning for sustainability in Mauritius.

The UM–IIASA group had three aims. Each is important, and provides an example of how scientific problem solving can also be set in the kind of practical problem-solving process IUCN has been promoting for its NSSD. The aims of the model and the modelling exercise were:

1. To help Mauritian decision makers formulate good sustainable development policies.
2. To address basic scientific questions of the links between population, development and the environment.
3. To reinvigorate the process of sustainable development modelling.

One of the first questions faced is: what is the environment? It might also be phrased as 'what is *not* the environment?' In one sense nothing can be excluded. On the other hand, some limitation and specification is absolutely necessary. In the frameworks above, we saw the environment defined in an ad hoc and highly limited manner as a specific impact, such as CO_2 emissions into the atmosphere. The Mauritius–IIASA group provides an elegant and practical solution by turning to the classics, and seeing the environment as the four elements of the world: earth, water, air, and fire.[8] If fire is seen as energy, this provides an effective way of dealing with what are today major environmental concerns. Population is seen at the centre of this system, rather than as something separate from the environment. The point is made that the environment cannot be

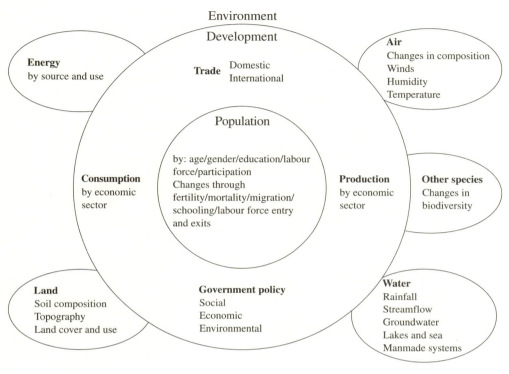

Figure 4 *IIASA's PDE model*

Source: Lutz, 1994

separated from all the species and subsystems for which it is the life support system. The basic structure of the general PDE model is shown in Figure 4.[9]

To turn this general idea into a specific model, the team followed three basic modelling principles. They can be called principles of location specificity, simplicity, and flexibility.

Location-specificity. The model should only include relevant aspects that could potentially make a difference in the case considered. That is, it should be specific to a given country or region, and not a general model to be applied in fixed fashion to all cases, and it should include only those conditions that are most relevant to that specific country or region.

Simplicity. The model should be as simple and straightforward as possible, avoiding 'magic' black boxes.

Flexibility. Only unambiguous and direct relationships should be hard wired. All other aspects should be treated as part of the soft model structure, as expressed by scenario settings. That is, only the known and direct relationship should be specified and fixed in the model. Most of the relationships should permit specification by the user, testing different scenarios. Many models specify a large set of the relationships, often on questionable assumptions, thus their conclusions are built in at the outset.

9 Language presents a real problem here. Lutz (1994, p 211) rejects the common depiction of population and environment as two boxes connected by arrows flowing in both directions. He argues that '...the environment must not be represented as a separate box because environment and laws of nature are everywhere and no line can be drawn around nature. Nothing is independent of the environment, including the human population which is part of nature and in all basic life-supporting functions dependent on the environment.' Nonetheless in describing the various modules, it is difficult to avoid the language of population and economic activities having an impact on the environment, as though the two were separable.

10 The book edited by Lutz on the PDE model in Mauritius does not include a diagram of the model and its connections. This illustration is taken from an IIASA brochure advertizing the book.

11 Multi-state population projection models were originally developed to deal simultane-ously with several populations in different territories which interact with one another, for example through migration. As Lutz explains, this type of model need not be limited to geographic conditions. Thus it was adapted to deal with people with different conditions – age, education and labour force partic-ipation (Lutz, 1994, p225).

The Model and its Uses

Figure 5 provides a simplified picture of the overall model developed for Mauritius.[10] The model developed has four major modules: for Population, Economy, Land Use and Water. In addition, there is a module for policies, which makes it especially useful for planners and decision makers. Each module has many sub-elements, and overall there are nearly 1000 parameters. One notes immediately that there is no module for air, one of the four elements of the environment with which the team started. This important omission illustrates the simplifi-cation principle (see above). In Mauritius the constant winds blow all atmospheric emissions out to sea, so that atmospheric emissions, or air quality, is not a problem. In other settings, an atmospheric module would be very important, especially for the connections between development, energy sources, air quality and human health.

Another specifically Mauritian adaptation is important. The model was developed for the entire island as one region or ecosystem. This is defensible in the case of the small island of less than 2000 square kilometres. Where movement of the population from one place to another is usually a matter of minutes or hours rather than days, such a decision is quite appropriate. For a large country, however, it could well make sense to develop both a national model and a series of regional models for important regions. Models can be developed for states, provinces or large metropolitan areas, and linked to a national model. In most cases it makes sense to develop one model for the nation as a whole, since nations do formulate policies and (at least attempt to) control their boundaries. The modules for the Mauritius PDE are briefly described below.

The Modules

The Population Module

The population module uses a multi-state population projection model that includes age, sex, education and labour force participation.[11] Its basic structure is shown in Figure 6. This represents an immense improvement over other frameworks and models of population and the environment, which often take population as numbers alone. Numbers imply seeing people only as consumers of resources. Distinguishing age and sex brings us to important differences in consumption among different groups of the population, and thus to different consumption patterns or levels for populations with different age and sex structures. Adding education and labour force brings to light the fact that people are not only consumers but producers as well, and that their productive capacities can vary, largely as a result of differences in education. Labour force participation rates will have a variety of important uses. When viewed within the population module alone, they will have an important relationship to fertility. High fertility limits women's participation in the labour force, declining fertility increases the available labour supply. When viewed in connection with the economic module, which estimates the demand for labour, it brings estimates of employment and unemployment, which are surely among the most important and explosive social issues in development planning. A society which experiences rapid economic development with high levels of unemployment is vulnerable to social unrest that can retard development. It can also signal especially heavy environmental degradation as the unemployed poor seek to eke out a meagre existence on vulnerable soils.

The population module can be run separately as an independent model. It is connected to the economy model by the output of labour distinguished by age, sex and education, and the input of consumption by age and education. In this module, Lutz develops a new twist on the familiar dependency ratio (see Part IV below – dependency ratio). The typical ratio is the Total Dependency Ratio, which is simply a function of the age structure. It is usually taken as the number of dependents (people under 15 and over 60) per 100 people of working age. For Mauritius, Lutz incorporates labour force participation, different consumption patterns for different dependent age groups (the aged consume more health services than the young, the young consume more education than adults not in the labour force, for example), and the productivity of the population due to differences in education. When all of these adjustments are included the new dependency ratio is called the Socio-Demographic Dependency Ratio (SDDR). This refinement of the dependency ratio does not make much difference to Mauritius's recent changes. All the various refinements have produced about the same, roughly 40 percent, decline in the dependency ratio. But future changes are likely to be quite different and closely tied to changes in labour force participation and education levels, both of which are subject to a variety of social policies that can be identified, and whose repercussions can be traced out in different future scenarios. Thus this refinement is likely to be quite important in future planning.

The Economic Module

The UM–IIASA team decided to use an 'input–output model for a small open economy' for Mauritius. Adhering to the simplicity and location specificity principles, the team developed a 15-sector input–output matrix. The choice here is between input–output and

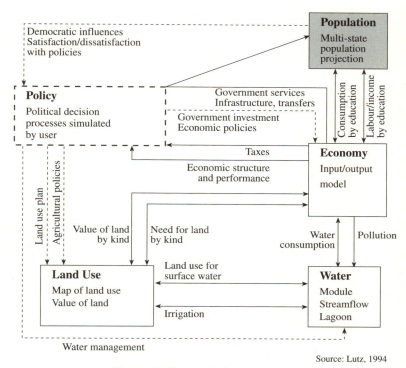

Source: Lutz, 1994

Figure 5 *The model for Mauritius*

general equilibrium models to depict the economy. Economists will debate the relative advantages and disadvantages, but it is possible that the input–output model will be more intuitively accessible to non-economists, because it begins with a set of commonly recognized industries or sectors.

The selection of sectors is closely tailored to the Mauritian economy. It includes sectors for sugar cane, sugar milling, Export Promotion Zone textiles and EPZ other, hotels and restaurants to capture the tourist industry, and a minimal standard set of sectors for electricity, water, construction, transportation, finance, and government. Export demand is exogenous, but there are different policies and assumptions about their impact, which can be

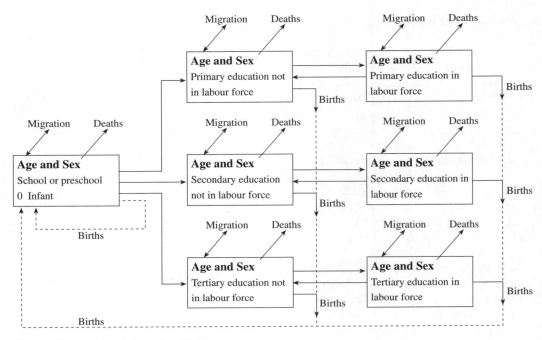

* The model for men is identical apart from 'Births'.

Source: Lutz, 1994

Figure 6 *Basic Structure of the Population Module for Women with Seven States**

specified in the scenarios. This reflects the principle of flexibility. The success or failure of Mauritius's export policies in not predetermined in the model.

The same can be said for government expenditure. It is paid for domestically, from government taxes and import duties, but the model does not require an automatic balanced budget. To balance the budget the scenario-maker must experiment with different tax and investment regimes and rerun the model several times. Alternately, if specified taxes are insufficient for a proposed investment programme, foreign borrowing is specified. In running these different scenarios and experimenting with different policies, the decision makers gain some insight into the dynamics of government finance.

The model is run together with the popula-

tion, land use and water modules, and there are some specified pairs of variables that, for realistic scenarios, either must have similar values, or only one of the two can be larger. Taxes and government consumption must be relatively equal – large deficits or surpluses will call for another scenario in which taxes or expenditures can be adjusted. Labour demand should not exceed supply. The opposite would be realistic, however, as it would indicate unemployment. A labour shortage would have to be adjusted for by increasing the productivity or participation rates. The model can estimate labour requirements by sector. Investment spending and savings must be equal. In all of these areas, the decision maker runs different scenarios and must often go back to make adjustments to produce the required balances, which reflect realistic scenarios. As the authors remark, 'In the course of this effort, the user will acquire an understanding of the interactions in the system. *This is probably as important as the results of the final scenarios themselves.*' (Lutz, 1994, page 248, emphasis added).

The Land Use Module

Like the economy module, the land use model is closely integrated with the population and water modules and cannot be run independently. It is connected to the economy model by the value of land and the demand for land by specific type. It is connected to the water model by the role that irrigation and surface freshwater flows play in agriculture and in the hydrological cycle. It is also connected to the policy module through agricultural policies and land use planning.

Again, simplicity rules and the land is divided into only five main categories: sugar, other agriculture, urban, beach, and 'other' non-commercial land. Because of its history and current political–economic structure, sugar is an important crop covering over 80,000 hectares or about 40 percent of the total area. Sugar owners are important political groups; growing and milling employ a substantial portion (17 percent) of the labour force and account for a substantial share of the export earnings (35 percent). The government has made a policy decision to maintain the sugar industry, especially since the Lomé Accords guarantee exports of roughly half a million tons to the EU under negotiated (protected) prices.

The most distinctive element of the Mauritian land use module is the sector for beaches. Some 125 kilometres of beaches are available for tourism, with a total area of just 12.5 square kilometres. This borders a series of lagoons enclosed by coral reefs, which break the heavy surf, preventing land erosion, and providing excellent sites for the tourist

industry. Although this is a very small area (just over one half of one percent of the total), it has a prominent place in the land use model because tourism has become a major industry, employer and export earner. Both the coral reefs and the enclosed lagoons raise the importance of the issue of water pollution, which is treated in the water module.

Over the past 30 years, sugar land has declined by about 15 percent, or 8 percent of the total land area. Urban land use had increased by 215 percent, to about 7 percent of the total area. The land use model does not change the amount of land to be used as beaches, but it does allow for changes in concentrations of hotels and tourists. Competition for land arises from powerful market forces, which show much higher value added by tourist than urban land, and much higher for urban land than for sugar, which has the lowest value added per square kilometre. Still, market forces will not be the only determinants of land allocation. Long term government policies will be as important, making the use of simulation modelling especially important.

The Water Model

Like the population module, the water module is sufficiently complex and independent for it to be run independently of the others. The water module is connected to the population model only through the economy module, with water consumption and pollution as the critical links. It is connected to the land module, as noted above, through irrigation and surface water use.

The hydrological model in the module shows an extensive system of 93 short rivers in 47 river basins, river diversions, man-made lakes, and underground aquifers. This supports a water intensive economy, with heavy irrigation for sugar and the water hungry textile industry. In addition, the high population density implies a large demand for domestic water use. The major water stocks and flows are mapped with good time series and seasonal data available for the modelling exercise.

Another distinctive aspect of the Mauritius PDE is the development of a 'lagoon model' as an element of the water module. This is necessitated by the importance of the tourist industry, and its vulnerability to pollution of the lagoons. Fast running rivers carry discharges from sugar, agriculture, industry and urban use into the lagoons, some with and some without treatment. Pollution is assessed by biological oxygen demand (BOD) and by nitrogen and phosphate loads, primarily from agriculture. There are also chemical effluents discharged by the textile producers, but little is known about them, and they are not now modelled. This may be a serious problem in the not too distant future, as might be the heavy loading of nitrogen. The model shows some limitation here, primarily from lack of data. First, only two pollutant measures, biological oxygen demand and the stream flow (in cubic metres per second) required for dilution, are tracked with different scenarios. Second, the lack of data precluded the development of regionally specific lagoon models, though it was well known that some lagoons receive much more pollution than others, and their degradation is already affecting the tourist industry. The lagoon model can only be run as a single global model. These two limitations are, however, the result of data deficiencies that can be corrected.

Water policy variables include quality standards, investment in treatment and investments in storage.

Scenarios

With these modules and the extensive data that were put into them both from existing sources and from the knowledge about specific linkages, the decision makers can run a great variety of scenarios. First, since data are available from 1962, the model can be run from 1962 to the present (or in the published study to 1987) both to validate it and adjust its calibration. Next one can ask different questions about the relation between population, development and the environment. These are of the 'what would have happened if...' variety tracing alternate histories. Finally, one can trace out some of the implications of different policy options or examine the impact of changes (such as a collapse in the world sugar market) in other conditions. We cannot review here all of the scenarios run in the published version, but we can provide a summary of some of the more important lessons, or questions answered by running the scenarios. There are both alternative histories, and future scenarios that can be examined.

It is important to keep in mind that in the brief span of 30 years, from 1960 to 1990, Mauritius was transformed from a typically poor, agrarian economy with high population growth and only moderate life expectancy to a modern, low fertility country with high levels of education and welfare. It also increased its level of wealth by ten times, with good prospects for further development.

Alternative Histories

1. What would Mauritius look like today without the demographic change and economic development?

• Population would be larger and poorer: 1.55 million in 1992 rather than

1.1 million; the school population would be twice as large (603,000 rather than 308,000). Per capita real income would be a third of what it is (6,000 rather than 18,000 rupees), and declining rather than growing. Unemployment would be at 20 percent rather than 9 percent; foreign trade and the budget deficit would have been unsustainable.

- Environmental pollution would have been less: stream flow requirements would have increased by 30 percent rather than by 60 percent (from 8.5 to 11.0 rather than to 13.4); BOD would not have changed appreciably, and would be less than half what it is today (0.09 rather than 0.23).
- Mauritius would still be a poor country, struggling to promote economic development under rapid population growth with a young and less educated population, and under inadequate public financial resources. Only the environment would have benefitted, though the population growth alone would have increased river water pollution.

2. What role did population dynamics play in Mauritius's development?

- Reduced fertility increased the supply of women to the labour force. Increased education helped reduce fertility and increase the productivity of labour. Both supported the economic boom that took off with government planning for both export promotion and tourism.
- A failure to reduce fertility alone would not have reduced per capita income, but it would have produced an unsustainable budget deficit (because of greater education and health costs), and increased water pollution 10 percent above current levels.

- Economic stagnation alone, with the fertility and mortality reduction that were actually experienced, would have produced a much lower level of (and declining) per capita real income, a high level of unemployment (36 percent), but substantially lower levels of water pollution.

Future Scenarios

Mauritius has experienced an enviable pattern of social and economic development over the past 30 years as a result of a variety of political, social and cultural conditions (see Box above) that portend well for the future. There is good reason to expect that the past quality of both leadership and citizenry will continue. A variety of future scenarios can be modelled, using a number of plausible assumptions. Most interesting, however, are the findings of a scenario that continues current policies, practices, and trends. This is called the *modern/boom/garden* scenario.

This scenario clearly points to an environmental crisis. The BOD increases from 0.27 kg per m^3 to 0.56 in 2020 and 0.93 in 2050. This raises the prospect of killing the corals and killing the lagoons through eutrophication. The result would be a collapse of the tourism industry. Any reasonable scenario for development is not sustainable because of increasing pollution.

- The solution is early water treatment. An early water treatment scenario reduces the stream flow demand to one-quarter of its current level, and the BOD is cut almost in half by 2020, and to one-quarter by 2050. The lagoons are saved, along with the tourist industry.
- The cost is high in the short run, though it soon declines to insignificance. A reasonable solution implies investments of 4.75

Mauritius: Critical Determinants of Success

Consider first some of the milestones in recent Mauritian demographic and economic history.

1948–9 Malaria eradication campaign drastically reduces mortality; fertility climbs to high levels. In 1950 CDR was 13.9, and CBR was 49.7, implying 3.6% growth, or doubling in 20 years.

1950 Parliamentary proposal to promote family planning meets with strong opposition from Catholic and Muslim populations. Parliament forms a representative committee to address the population issue.

1953 Committee recommends government promotion of social services including family planning. Catholic leadership accepts family planning using natural methods (periodic abstinence).

1957 Mauritian Family Planning Association is formed with assistance from the International Planned Parenthood Federation.

1960s Government promotes import substitution through a Development Certificate scheme.

1961–2 Two reports on population (Titmuss) and economic conditions (Meade) note need for combined fertility limitation and promotion of economic development.

1962 Catholic Association Familiae formed (to promote periodic abstinence).

1967 Mauritius achieves independence under fully elected government. Socialist government forms coalition with the conservative party representing sugar planting interests.

1970s Government shifts to export promotion policies, forms Export Promotion Zones, textile industries grow significantly.

1980s University and government collaboration in the Mauritius 2000 project, drawing in UNFPA and IIASA to produce the Population–Development–Environment model for Mauritius.

What explains the positive and progressive development? The potential for the population to overshoot its resources was great, but it did not overshoot. We can see the convergence of a number of conditions that made the adjustment and progress possible.

1. Ecological pressure. The conquest of malaria using new insecticides brought a rapid reduction of mortality and population growth rates to unprecedented heights, generating concern in all sectors. This was exacerbated by the island status, giving it no new frontiers to open, and by the economic conditions that produced high levels of unemployment in the 1950s. The signals of danger were clear.

2. Social capacities. The island's population was well educated, even in this early period of relative poverty and rapidly increasing ecological pressures. The mixed ethnic population had a history of mutual acceptance, rendered easier by the fact that none of the ethnic groups was indigenous to the island. Social capacity to respond to signals was high.

3. Leadership. A tradition of responsible leadership had developed, encouraged by the smooth transition from colonial to independent status. Political leaders moved to share power and to build consensus across both ethnic and class lines. Religious leadership was also crucial, especially in the person of the Jesuit Professor Lestapis, who promoted religiously acceptable forms of birth control, thus precluding an unresolvable religious debate over the population issue. More high social capacity to respond to signals.

4. Democratic forces. The combination of a well educated population and a leadership promoting consensus and social services has kept Mauritius a fully democratic nation. Government is responsive to citizen interests and demands, and the citizenry is responsible and loyal. Response to signals was likely to be effective.

billion rupees per year. This would be 21 percent of GDP in 1990, but would fall to 4 percent by 2020 and to only 1 percent by 2050.

Three of the interesting findings here relate to population, economy and the environment.

First, fertility reduction in Mauritius was clearly important in many ways, but it was less important than economic development in producing the current high levels of quality of life that the Mauritians enjoy.

Second, the environmental impact provides one answer to the question of the relation between development and degradation. Mauritius's past economic development did indeed increase water pollution, and the past processes are clearly unsustainable. The modelling experiments show that continuing current practices would produce an overshoot in lagoon pollution, with probably drastic repercussions on the high earning tourist industry.

Third, at the same time, the future scenarios show that overshoot can be avoided; the pollution can be reversed. It will take substantial political will and good economic planning, plus most likely some extensive foreign assistance (both grants and loans) to correct the process, but it can be done. Thus sustainable development for Mauritius is a clear possibility, and the modelling tool helps to point the way. Sustainable development for Mauritius will require adjustments to unsustainable practices. If these are made, it will imply a continuing rise in the wellbeing of both people and ecosystems. It is also clear that the alternatives, with no heavy investment in water treatment, could well mean a crisis or collapse of certain activities based on a protected environment.

Modelling for Mauritius helps make clear the possible future scenarios, and suggests ways that the more favourable scenarios can be achieved. This makes modelling a highly important, possibly indispensable tool for promoting sustainable development.

Next Steps

12 As noted above, the categories 'indigenous' and 'non-indigenous' are social definitions, and do not necessarily bear close relationship to the actual histories of human migrations. How long it takes to become indigenous, and who gets included in the category, are social definitions, not necessarily historically justified. That does not make the differences less important or the conflicts less serious. It merely reminds us that these social definitions are powerful determinants of mobilization and behaviour in themselves.

13 The fate of the ECCO model provides support for the advantages of simplicity. Originally it was planned to develop and run the ECCO model on Mauritius. Even the simplicity of Mauritius could not help the ECCO model, however, and the plan was apparently abandoned. It might be that ECCO will ultimately prove useful, for its use of energy units rather than prices. For the moment, however, it must be judged too inaccessible to have much value.

People who are sceptical of the Mauritius modelling exercises have pointed out rightly that the specific case may have little relevance for sustainable development planning in general. The problem lies in the simplicity of Mauritius. It is a small island state, with a relatively simple set of conditions. Air pollution is no problem, the full economy can be captured quite well with 15 sectors, and the entire country can be treated as one region. Even land use is sufficiently simple to be captured by merely five categories. Less often recognized by sceptics, but probably of great importance, is the lack of an indigenous population, which precludes the conflict between indigenous and non-indigenous peoples that we see in so many areas.[12] Another type of simplicity arises from the open, shared character of political leadership on the island. That the new popularly elected leaders chose to share power with the wealthy sugar interests greatly reduced the potential for class conflicts that would have complicated the situation. For all these reasons, sceptics have argued that the Mauritian experience is not really relevant to other cases, where there are far greater complexities.

In fact, one of the reasons Mauritius was chosen for this exercise was its simplicity. The IIASA team wanted to reinvigorate systems modelling as a tool for examining population–environment relationships, and for making those relationships adaptable to national level planning. For this it was most useful to begin with a very simple situation, which could be grasped relatively easily.

There are two reasons to reject the scepticism, however. One is strategic, the other empirical. Strategically, the Mauritius experience demonstrates that systems modelling can be done for an entire economy in which the population–environment relationship is critical and central to the exercise. In this respect, it is probably impossible to overstate the importance of the decision to develop and use a multi-state population projection model for the population module. This was a decision made by a well trained demographer, Lutz, and was based on the recent development of the specialized technique of multi-state projection models, which were then only common in the literature for less than a decade. It is always useful to be able to demonstrate the power of a new tool, and it was probably easier to make this demonstration in a simple, rather than a more complex, situation. Had a more complex situation been chosen, even this important tool might have been lost if the situation had proven too complex to be modelled successfully.[13]

Empirically, the IIASA model is now moving onto other and more complex situations. A student of Lutz, Ms. Anna Wils, undertook a PDE analysis of Cape Verde (Wills, 1995). This is another relatively simple situation, of course, though it is much poorer than Mauritius, in both resources and data. It is plagued by past poverty, colonial and land tenure policies that discouraged increases in productivity, and a chronic shortage of water that kept both populations and output in check. In this poor land, emigration and remittances have kept the standard of living from falling below the poverty line. Nonetheless, the PDE model could show varieties for future scenarios. One suggests what would need to be done to move in the direction of Mauritius's 'garden' type of economy. Another shows what will happen to population, quality of life and quality of the environment under a stagnating scenario.

More important, however, is the new exercise being conducted in the three Yucatan Peninsula states of Mexico – Campeche, Quintana Roo and Yucatan. In this case IIASA will work with a Mexican research organization, the Merida Unit of CINVESTAV, an organization founded in 1980 to promote social

and economic development in the region. This is a far more complex situation, with a history of conflict between indigenous groups and large landowners, complex ecosystems, and complex connections to both the state of Mexico and the larger world community. The study will also reach further back in time to try to understand the collapse of the Maya civilization, and deeper into the rich micro-ethnography of the area. Thus it promises to develop new techniques for population–environment analysis at the same time that it shows the utility of the PDE model for regional level planning for sustainable development.

Part IV

Population Parameters and Dynamics

This section provides a discussion of population conditions that are especially relevant for issues of sustainability. It proposes the formation of a National Population Environment Review as part of the overall strategy for sustainability, and indicates a number of specific conditions that should be included in that review.

It begins with a discussion of the demographic transition, noting that there were two, not one, and that the policy implications of this are profound. They show that:

Population conditions can be changed more easily with less human cost today than in the past.

The section then presents a series of suggestions for measuring and mapping a variety of ecological conditions – population, environment, technological and organizational – that should be included in a National Population Environment Review.

The Demographic Transitions

The demographic transition – the movement of a population from high to low mortality and fertility – is one of the best known demographic observations (Notestein, 1945). It is not a theory, though many of its aspects are covered with theoretical dispute. It is an empirical observation, and a near universal one at that.

There is unfortunately a common misperception that the demographic transition is a single and rather fixed transition. It is most important to note, however, that there are not one, but two demographic transitions, past and present, (Figure 7) and the policy and programmatic implications of this distinction are profound.

If rapid population growth is a problem, which most people today agree it is, it is a problem that can be addressed. A combination of measures can be taken that can reduce fertility rapidly and raise the quality of life. These include education for women, effective primary health care especially for women and children in rural areas, high quality, client-centred family planning services, and the promotion of women's rights. It is largely the responsibility of governments to provide this range of basic social services. When they do, and when they act effectively, there are benefits for the environment, for economic development, and for individual welfare, especially for poor rural women and children. When governments do not provide these services, there are high costs for the environment and for popular welfare, especially for poor rural women and children.

But to see clearly the role of social services and government responsibility, it is necessary to be clear about the character of the two demographic transitions. Figure 7 portrays some of this character.

The past demographic transition took place in all of the currently industrialized countries, and the same broad pattern, with somewhat varying rates of change, occurred in all the industrialized countries. Birth and death rates declined from high (or natural) to low (or controlled). Mortality declined first, rather gradually, followed, after a period of up to a century, by a decline of fertility. The intervening period witnessed relatively high rates of population growth. Those rates slowed with the decline of fertility and now every industrialized country shows fertility at or below replacement levels (a total fertility rate of 2.1 – see below, p. 122).[14] With these rates all will shortly begin to, and some already do, show negative population growth rates.

All currently industrialized countries went through this transition. There were differences

14 Industrial or high income countries today (1990–95) with Total Fertility Rates above 2.10 include the Republic of Moldova (2.13) and Iceland (2.23) in Europe, and New Zealand (2.17) (UN, 1994). See below for a discussion of the total fertility rate and replacement level.

> The demographic transition is the movement of a population from high to low birth and death rates.
>
> There are two demographic transitions, past and present.
>
> The policy implications of this are profound:
>
> Population conditions, especially human fertility, can now be changed more easily and at lower human cost than in the past.

Past demographic transition (England and Wales, 1700–2000)   Present demographic transition (Less developed regions)

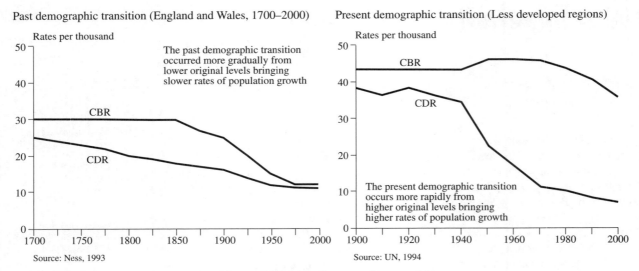

Figure 7 *The two demographic transitions*

in the timing and trajectories of both mortality and fertility declines, and in the rate of population growth in the intervening years. But all went from high to low mortality and fertility. In the process they all moved roughly from rural–agrarian to urban–industrial societies.

The present demographic transition is what we find in process throughout the low income or less developed regions of the world today (Figure 7 uses data for the UN's category of less developed regions). In some countries the full transition has already been completed (China, Korea, Taiwan, Hong Kong, Singapore, Thailand, Barbados, Cuba). In others, mortality has fallen and the fertility decline is well advanced, but levels are still above replacement level (Indonesia, much of Southern India, Sri Lanka, Tunisia, Brazil, Chile, Colombia, Costa Rica, Panama, Uruguay). In still others, mortality has fallen but fertility remains high (most of Africa, Pakistan, and the Middle East).

Although there have been many differences in both past and present demographic transitions, it is possible to make a broad generalization that is of considerable significance. The past demographic transition took place gradually, from comparatively low original levels of both mortality and fertility. Thus the intervening period of rapid population growth moved slowly, with rates that seldom rose above 1 percent per year. The present transition starts from higher levels of mortality and fertility (for reasons that are not at all clear), and mortality declines come much more rapidly, leading to growth rates of 3 percent and more. With the right conditions, fertility can also fall very rapidly, as we have seen in China and Thailand.

Many of the differences – between past and present transitions, and among countries in the present transition – are based on differences in technology, and in social policy.

Technology

Past declines in mortality and fertility took place with almost no important technological advances in medicine. Progress in public health was important, especially in providing clean water and safe sewage disposal. But medicine as such played little part. The only significant medical advance at this time was the discovery of a safe smallpox vaccine at the end of the 18th century. It was not widely used, however, until the 20th century.

Mortality declined from a gradual rise in the standard of living. This came about through revolutions in world wide trade, agriculture and industrial production initiated in the 'first world', and closely associated with the use of fossil fuels. There was also some 'global warming' as Europe and China came out of 'the little ice age' in the 18th century. In effect, modern technology for production and transportation, and urban–industrial social organization increased the carrying capacity of the earth.

Fertility declined with the rise of urban–industrial society. With this transition, children became more a liability than an asset and people desired fewer (Caldwell, 1976, Livi-Bacci, 1989). Moreover, advances in education, especially for women, raised the age of marriage, leading to lowered fertility. The fertility limiting technology remained essentially the same as had been practised since the beginning of human history: abstinence, withdrawal, and abortion, with infanticide added when fertility limitation did not work. Throughout history, abortion has been one of the most common means of controlling fertility (Livi-Bacci, 1989, Jacobson, 1994).

Present declines in both mortality and fertility are in large part driven by major technological revolutions in medicine, which are largely a product of the middle of the 20th century. Antibiotic drugs, new vaccines, pesticides and fungicides were vastly improved, in part due to the second world war, and came to be widely used to control infectious diseases in the immediate post war period. This brought revolutionary declines in mortality. Declines that previously took one or two centuries in the industrialized countries now often took place in one or two decades in developing regions.

The new contraceptive technology is even younger. It only became generally available in the 1960s. The intra-uterine device (IUD), oral contraceptive pills, injections, subdermal implants, and vastly improved, safer and cheaper procedures for sterilization and abortion constitute a technological change that can only be called revolutionary. This technology has permitted exceptionally rapid declines in fertility. As with mortality, fertility declines that in the past demographic transition took one or two generations, are occurring today in one or two decades.

It is only dimly perceived in some areas, and very much resisted in others, that this revolution is especially important for women and for gender roles. For the first time in history, women can have substantial control over their fertility.

Perhaps the most important characteristic of both the mortality and fertility controlling technologies is that they are what can be called 'bureaucratically portable' (Ness and Ando, 1984). The technology can be placed in large scale bureaucratic organizations and distributed and applied in a highly routinized fashion. In some cases, as for example with water purification, the technology may not even require the acceptance of the masses of users. Insecticide applications can be carried out on a wide scale with almost military organization, and even imposed on individual houses by government decree. Children can be inoculated in schools or rural health centres. The smallpox vaccine was developed as a new technology at the end of the 18th century, but moved very slowly for the next century and a half. When its distribution became the responsibility of the WHO and

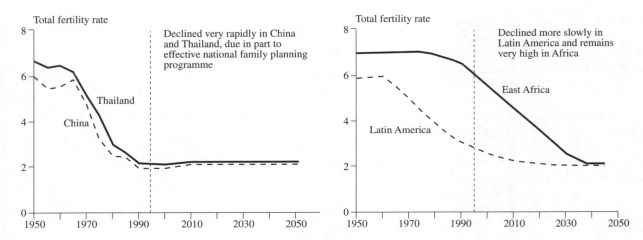

Figure 8 *The total fertility rate*

national government health departments, smallpox was effectively eliminated in about three decades. We have seen widespread campaigns for the control of infectious diseases carried out with great success world wide, largely by using large scale bureaucratic organizations to develop delivery systems for the application of the new technology.

The rapid declines we have seen in fertility are also associated with an effective delivery system for the new contraceptive technology. Where it is developed and open, the market itself can provide an effective delivery system, as it does in most of the more developed regions. In the less developed regions, there has also been the rapid, and somewhat revolu-

tionary, development of national family planning programmes. These may be government programmes, as in most of Asia, or private non-governmental programmes, which are more prominent in Latin America. In some cases they are highly specialized in fertility limitation alone, and in other cases are closely allied with health or other social services. There are many types of organizations to be found, but the successful ones tend to show similar characteristics. They must be widespread, operate on a large scale, and bring roughly similar services and supplies to widely dispersed settings. Without such large scale, bureaucratic-type organizations, providing the new technology to a large population would be very difficult. The importance of organization implies the importance of policy change.

The Anti-Natalist Policy Revolution

Past fertility transitions took place in countries where governments were often pro-natalist, and openly hostile to proponents of birth control.

Present transitions are taking place under governments that promote fertility limitation within marriage to limit rates of population growth.

Policy and Policy Change

In population policy we are witness to one of the most profound, and quiet, of policy revolutions. It has been called the anti-natalist policy revolution (Ness and Ando, 1984). Until 1952 almost all governments had been pro-natalist.

Technology, Organization, and Policy in China's Fertility Decline

China's fertility decline ranks as one of the world's fastest. From a total fertility rate of 5.94 in 1965–70, it declined to 1.95 in 1990–95 (UN, 1994). China has been criticized for the often coercive nature of its national family planning programme, but the importance of both the technology and the previous organizational development are often neglected. From the early 1950s, China embarked on a massive campaign to provide primary health care to all the population, and especially to the rural areas. More than a million 'barefoot doctors' were trained and provided with simple medicines and procedures to help curb infectious diseases. This established an extensive government delivery system that proved to be highly successful. In 25 years it brought the infant mortality rate from 195 to 50. When China finally adopted a strong anti-natalist policy in the early 1970s, it could use the extensive delivery system it had developed to provide information, service and supplies for contraception. It is difficult to see how China could have achieved such a rapid rate of fertility decline without the new contraceptive technology, or without the delivery system it had developed prior to making its important anti-natalist policy decision.

When India attempted to adopt a more coercive fertility limitation programme in 1975–77, without the type of national political organization or extensive primary health care system China had, the result was exceptionally costly both for the government and for the fertility limitation programme that the government had adopted much earlier.

Thailand parallels China in the rapid decline of both infant mortality and fertility. Like China, it extended its primary health care system, especially to the rural areas, and decided in 1970 to include family planning along with such care. Thailand shows that such progress can be made without coercion, if the technology is used and an effective organization is created for its delivery.

The well known policies of pre war Germany, Italy and France are only prominent examples of what has been pervasive for centuries. Margaret Sanger was jailed in America, and Mrs Kato in Japan, for advocating birth control in the 1930s. All governments have seen people as a resource: to be taxed, worked and sent off to war. All governments have wanted more rather than fewer people, and have intuitively seen that the best way to have more is to encourage people to act naturally, to reproduce. Thus past governments have typically been pro-natalist.

All of that began to change in 1952, when India became the first country to announce an official policy to limit population growth by limiting fertility within marriage.[15] There are two revolutionary aspects of this change. One lies in the new aim to limit population growth. The second, which may be even more revolutionary, is the attempt to reduce population growth by limiting fertility within marriage. In all human societies the marital bond carries with it the right and the obligation to reproduce. Now for the first time, the state is attempting to intervene directly and deliberately to reduce fertility within that bond.

Pakistan became the second country to adopt an anti-natalist policy, in 1960. Since then almost all of the low income countries have adopted some form of anti-natalist policy. More than 90 percent of the population in the low income countries is ruled by governments that have made this policy transition.

15 Japan also adopted a fertility limitation policy in 1952, the culmination of five years of increasingly liberal abortion law changes. Japan's policy was, however, aimed at increasing female health more than at reducing population growth. The great social and economic stresses of the post war period led many women to resort to abortion. Unsafe abortions became a serious health hazard to which the government reacted by making abortion legal and safe. Note that this was one of the few options Japan had, since an effective contraceptive technology was still more than a decade away.

It is useful for scientific analysis, and thus for a better understanding of policy, that the strength of this policy varies and can be measured (Ness and Ando, 1984). A strong policy typically implies a broad attention to the causes of fertility, focusing on both social service provision, attempting to raise the age of marriage, and making available the full range of the modern contraceptive technology. There are very strong policies in such countries as China, Korea, Thailand, Indonesia, and Sri Lanka. Policies are moderate in India, Bangladesh, most of Latin America and in Tunisia, Botswana, Kenya and Zimbabwe. Policies are weak in Pakistan, and most of Africa. And in some countries, such as Burma and Saudi Arabia, there is no policy change; the countries remains pro-natalist. The strength also varies over time, with some countries adopting the policy earlier than others, and some (as Pakistan) showing considerable movement back and forth over time.

Causes

These policy changes are driven by a complex set of conditions, whose individual weights vary depending on local political and cultural conditions. Identifying them can help us to understand what governments can do to promote the kind of population policy that supports sustainable development.

We can identify major conditions that have caused the recent policy changes:

- The strength of the political administrative system;
- government commitment to economic development;
- population density;
- a history of population census and social data collection; and
- contraceptive technology.

See, above, page 26 for fuller discussion of the Political Administrative System

From the history of policy changes, it is easy to identify how each of these conditions has affected the change. The conditions also mark the current differences between weaker and stronger policy positions.

Countries with strong political–administrative systems tend to have stronger policies. A strong system implies that governments are trusted and can mobilize popular sentiments and action. They can generate consensus and action, even when resistance is found.

Countries with a strong commitment to economic development tend to have strong policies. The connection is very direct here. Commitment to development leads to economic planning and data gathering, which show very clearly to finance, health and education ministers the high costs of rapid population growth.

Population density is associated with earlier and stronger policies, though the theory explaining this connection is not yet well developed. Some evidence (Zhang, 1994) suggests that high density leads policy makers to perceive that population growth presents an unsustainable pressure on the society. There are more direct and less costly ways to help policy makers see the implications of rapid population growth, including computer models with effective graphic capacities (Lutz, 1994, Futures, nd).

Data collection is also associated with policy strength. Asia's round of 1960 censuses had a strong impact on policy. Africa's lack of almost any census data until the 1980s delayed policy changes. In addition, the use of social and economic surveys showed governments both the costs of population growth, and the willingness of people to limit their fertility.

Finally, the new contraceptive technology has helped greatly, since governments who wish to reduce population growth rates now have a clear instrument to achieve this end. For example, Chinese leaders saw and were concerned about the high rate of population growth as early as 1953, but recognized they did not have the capacity to do anything about it. (Ness and Ando, 1984).

Consequences

One of the most important consequences of this policy change lies in the area of implementation, which usually takes the form of a national family planning programme. The first of these explicit, specialized, large scale organizations was established in India in 1952.[16] Today there are more than 100, both governmental and non-governmental, throughout the low income countries. Much is known about how these programmes work, and as in policy, the strength of programmes varies and can be measured (Ross, Mauldin and Miller, 1993). The national programmes of nearly 100 countries have been assessed systematically by scholars at the Population Council, using expert judgement techniques. The measures of programme strength deal with what can be considered common sense conditions. Does the political leadership give strong vocal support? Are clinics and service personnel well distributed and close to the people they serve? Are the service units well staffed, well equipped and well supplied? Is there effective evaluation and monitoring? It is especially important to note that the measure of family planning programme strength is not closely related to a country's level of wealth or economic development. There are strong programmes in poor and in wealthy countries, and there are also weak programmes in poor and wealthy countries.

There have also been scores of national sample surveys of human fertility, in what has come to be one of the world's largest social science projects. The International Statistical Association undertook a series of national studies under the World Fertility Survey project starting in the early 1970s (Clelland and Scott, 1987, Phillips and Ross, 1992). These were followed by the more broadly-based Demographic and Health Survey organized by the Westinghouse Corporation under funding by USAID. In addition, there have been scores of smaller studies and operations research projects asking how family planning programmes work. Much of this is summarized in a recent review by the World Bank and UNFPA (World Bank, 1993; Sadik, 1991).

There has been much scientific controversy over the impact of family planning programmes since they gained momentum in the 1960s. Kingsley Davis (1967), and Mamdani (1972), provided some of the earliest arguments and evidence that the mere distribution of contraceptives has little impact on fertility as long as social and economic conditions are such that people want more children. Since then a substantial number of scholars have maintained the position that the desire for children explains most of actual fertility, and that family planning programmes can do little to alter the desire, since it is based on a rational choice that children are a greater asset than a liability. The most recent contributions to this debate (Abernathy, 1993 and Pritchett, 1994) provide evidence indicating that most of actual human fertility is determined by people's desire for children.

On the other side are scholars who argue that an effective family planning programme can supply information and persuasion that helps to reduce desired family size. Perhaps more important is the argument that for many people today, especially the rural poor, or people with inadequate access to basic health and social services, the number of actual births is greater than the desired number. Though abortions are more difficult to count, it is clear they are increasing in many countries, which causes

16 The programme went very slowly for the first two five-year plans, when it assumed its population growth rates would be 1.25 percent. The programme received only a small allocation, and did not use all the funds allocated. The 1961 census showed population growth rates to be over 2 percent, which led the government to double the programme allocation every year for the next five years, and to increase the effort put into the programme.

serious health problems, especially for poor women. This is often cited as evidence that people are having more children than they want (Ness, 1994), and is called the un-met need for fertility control (Sadik, 1991, Mazur, 1994).

Whatever the scientific outcome of the debate, it is becoming clear that the new contraceptive technology, together with the new organization of client-oriented family planning programmes, can reduce fertility far more rapidly, and with far less human cost than was possible in the past. There is no doubt that the strength of the family planning programme (as measured by Mauldin and Ross) can have an important and independent impact on human fertility and the quality of life. There are two important lessons from this review:

1. Programme strength is not related to the level of national wealth or economic development. Strong and weak programmes are found at every level of economic development.
2. Programme strength is closely associated with contraceptive use and fertility decline. At every level of social and economic development, programme strength is associated with higher levels of contraceptive use and fertility decline, and higher levels of maternal and child health.

These issues are treated at greater length below, page 106.

Environmental Implications: Past and Present

There is no doubt that past demographic transitions took place along with substantial environmental degradation. Since the 18th century, deforestation, destruction of habitat, and species loss throughout the world accompanied the transformation to urban–industrial society. There is little question that current industrial society, based on fossil fuels and chemical alterations of the landscape, is unsustainable.

Population growth has been one element of this environmental destruction, both past and present. But population growth cannot be separated from the technological and social changes of the transformation to urban–industrial society. There remain, however, differences between the environmental impact of the past and present transitions. Much of this has to do with the speed of the transition, and especially of the technological changes that accompany the transitions.

Past

The more gradual changes of the past probably permitted easier adjustment to the radical changes. Two examples must suffice. In Southeast Asia rice and rubber expanded rapidly from roughly 1850 to 1930. Tropical rain forests were transformed to paddy lands and rubber plantations. But this implied an expansion of wetlands (rice) and of forests (plantations) with technologies that were of long standing. Rice, for example, remained at its traditional 1–2 metric tonnes per hectare throughout this expansion. And rubber plantations without chemical inputs sustained a substantial portion of the traditional species. Moreover, much of the population growth during this period came from immigration, implying that the local societies did not have to support a young population through years of high consumption and relatively low production. In addition, standards of publicly provided social services were relatively low, implying that the costs to the societies were relatively low.

North American forests were devastated as the new country was opened up by massive demands for lumber for housing. Many areas

Caring for the Earth – on Population

The Earth is an Entire Ecosystem

The Earth's life support systems are the ecological processes that shape climate, cleanse air and water, regulate water flows, recycle essential elements, create and regenerate the soil and keep the planet fit for life. Human activities are radically altering these processes through global pollution and the destruction or modification of the ecosystem.

The unprecedented growth of human numbers and the expansion of human resource consumption places the entire planet at risk and is gambling with its survival.

Future Needs: Addressing Population Growth and Human Consumption

The Earth has limits. To live within those limits two things will need to be done: population growth must stop everywhere, and the rich must stabilize and in some cases reduce their consumption of resources.

This double approach to population growth and resource consumption is central to the IUCN position on population.

Development is Necessary

The aim of economic development is to improve the quality of life. Economic growth is necessary to improve the quality of life, but it is not an end in itself, and it cannot continue indefinitely.

Inequity is Unsustainable

Development and life quality are not evenly distributed.

A minority of the Earth's population enjoys a high standard of living and consumes more than its share of resources.

The great majority of the Earth's people suffer poverty and consume less than their share of resources.

Population growth rates are highest where poverty is most intense. Lack of social infrastructure and effective delivery systems are among the things that keep birth rates high and link population growth to poverty.

This inequity is unstable and unsustainable. Gross disparities in consumption and population growth rates must be overcome (p 44).

The Special Position and Needs of Women and Child Survival

Women are important managers of human and natural resources. Yet in most countries women have limited access to and control over income, credit, land, education, training, health care and information; and they suffer the worst effects of poverty and environmental degradation.

Ways must be found to correct traditional biases in the household against girls and women. Higher female literacy, cleaner water, better sanitation, availability of rehydration therapy and broader immunization are all linked to higher rates of child survival and in turn to population stabilization (p. 24).

Women must be given access to the means of controlling their own fertility and the size of their families (p. 23).

Effective and safe birth control can be achieved only if it is linked to improvement in the provision of health services, especially to poor people.

SPECIFIC ACTION RECOMMENDATIONS ON POPULATION AND SUSTAINABLE DEVELOPMENT

1. Increase awareness about the need to stabilize resource consumption and population.
2. Integrate resource consumption issues and population issues in national development plans and policies.
3. Develop, test and adopt resource-efficient methods and technologies.
4. Tax energy and other resources in high consumption countries.
5. Encourage 'green consumer' movements.
6. Improve maternal and child health care.
7. Double family planing services.

Source: IUCN, 1991

were clear cut for farmland, but others were simply left to revert to whatever would grow. After a century and more of this devastation, national governments began to create organizations to protect and regenerate forests. Today, total forest area is actually expanding in North America. As in Southeast Asia, much of the population growth in North America was made up of adult immigrants. Other societies had paid for the basic child care and education of these people, essentially providing substantial human capital investments from abroad.

In both cases, the immigration stream that contributed to rapid population growth also represented an important emigration stream that reduced the costs and pressures of those societies whose rapid population growth led to emigration.

Present

The present population growth is much more rapid than the past and much more costly for those experiencing the growth. There is less time for both populations and environments to adjust to one another. There is less opportunity to relieve pressure through immigration. Moreover, poverty implies a more static technology, or less resources for technological development, and rapid population growth with a static technology is almost inevitably environmentally destructive. At the same time,

the rapid expansion of production with new chemically based technologies is proving immensely destructive of the environment.

In effect, the slower growth of population and production with older and more traditional technologies may have been less environmentally destructive than is the current more rapid growth with a more radically transformed technology. Again, however, it is the combined impact of population, technology and social organization that has the major impact on the environment, and not population growth alone.

Even in the present, we can have rapid population growth along with environmental protection and even enhancement. Curitiba in Brazil (Rabinovitch and Hoehn, 1995) is a good example, as are cities in Japan and the US, and Singapore. In all of these cases population growth varied from rapid to slow, but all have managed to decrease air, water and land pollution in their immediate areas. This required new technologies and new organization (e.g. laws and government regulatory bodies) for environmental protection. The point of these examples is that rapid population growth can only be environmentally neutral or enhancing with a rapid advance in technology. With a continuation of a traditional slash and burn agriculture, for example, rapid population growth surely destroys the environment. We find this pressure of population growth on tropical rain forests throughout the equatorial belt.

A National Population–Environment Review

Caring for the Earth and *Agenda 21* recognize that information on population and its environmental impact will be basic to the process of integrating population in strategies for sustainable development. Such an integration can begin with a review of basic population conditions. Much of the data for such reviews are normally included in national development plans, and are also typically generated in population censuses or in demographic surveys. What is needed for effective integration of population in sustainable development planning, however, goes considerably beyond what is usually included in the demographic component of national development plans. Such integration calls for a review that examines specifically the links between population and the environment, or a national population environment review (NPER).

Six Classic Social Demographic Variables

The NPER can begin with a fairly standard set of population conditions. Firstly, there are the six classical variables from social demography.

One of the things that makes social demography an exceptionally powerful discipline is its limited focus on six variables. Population size, age/sex distribution, and rural–urban distribution constitute three fundamental comparative static variables. In addition, the rates of births, deaths, and migration constitute three dynamic variables. These six variables can be examined in a relatively closed system, or one that makes

simple assumptions of the external conditions that affect the rates. In effect, typical assumptions merely reflect historical conditions, or conditions currently found in different parts of the world, each of which is reflected in a standard life table. With such assumptions, it is possible to make projections of population size

---

Caring for the Earth and *Agenda 21*

Agenda 21 echoed IUCN's *Caring for the Earth* in its brief Chapter 5. That chapter made three basic recommendations.

a. Develop and disseminate knowledge concerning the links between demographic trends and factors and sustainable development.
b. Formulate integrated national policies for environment and development, taking into account demographic factors and trends.
c. Implement integrated environment and development programmes at the local level, taking into account demographic trends and factors.

As the basis for action for the second recommendation, *Agenda 21* notes that development plans in the past have recognized demographic factors, but in the future more attention will have to be given to these issues in policy formation and the design of development plans. To do this, it says, '...countries will have to improve their own capacities to assess the environment and development implications of their demographic trends and factors...They should combine environmental concerns and population issues with a holistic view of development, which includes: alleviation of poverty; secure livelihoods; good health; quality of life; improvement of status and income of women and their access to schooling and training, as well as fulfilment of their personal aspirations; and empowerment of individuals and communities.'

Six Basic Demographic Variables

Dynamic Conditions:	Comparative Static Conditions:
Fertility	Size
Mortality	Age/sex composition
Migration	Rural/urban distributions

and age/sex distributions for any point in the future. As noted below (Box, p. 116), these projections can be highly precise, but their precision does not necessarily indicate accuracy.

Technology and Human Organizational Variables

As we have indicated above, not only the number of people, but the tools they use and the way they are organized affect the environment. People with axes and oxen are less destructive of a forest than are people with chainsaws and bulldozers. Fishermen in small boats with lines and small nets deplete fish stocks less than do massive trawlers that work like a huge vacuum cleaner in the seas. The difference in technology is clear. But there are also differences in organization. A forest owned privately may be clear cut, next to a full standing forest protected as a national park or area. International agreements or national policies can limit net sizes and catches on the seas, or the movement and control of water. The Great Lakes of the US and Canada experience a reduction in pollution run-offs and depositions while the Aral Sea shrinks due to diversion of water to irrigate foreign exchange earning crops. Here the tools are the same, but the organization is different.

The NPER should include some indicators of these non-demographic conditions. Basic to these will be measures of wealth and welfare. Total wealth, sectoral output and rates of saving and investment are important, but are usually well covered in national development plans. These should also be linked to population and environmental conditions. Human welfare is commonly, and effectively, assessed by such measures as infant, child or maternal mortality rates. Literacy, school enrolment, or levels of education achieved are also commonly used measures, which speak both of welfare and of human capital, or the quality of the population.

Socially Defined Groups

In countries with mixed ethnic or other social groups, it is important to include such variables as religion, language, race, or ethnicity. These indicate social definitions, or differences between people that appear important to others. Such definitions are powerful mechanisms for both mobilizing and separating people, sometimes in violent ways. They define how groups will treat one another and are often associated with major differences in behaviour: in wealth, residence, occupation, reproduction, voting, and access to public goods. Race, or black and white, constitutes an important social definition in the US. The list is endless and would include Serbs, Croats and Muslims in the former Yugoslavia; Sinhalese and Tamils in Sri Lanka; Indians, Mestizos and Castilians in much of Latin America; and Tutsies and Hutus in central Africa. Not all of these common social markers are important everywhere. Christian–Buddhist distinctions in Korea appear to be relatively unimportant. Moreover, the importance of social identities can vary over time. Some remain powerful through many historical changes, while others, such as Roman Catholic and Protestant differences in Europe or North America, have considerably weakened in importance. There is today far less difference between Catholics and Protestants than there was in the past. Societies also vary considerably in their capacity to assimilate different groups. Chinese in Thailand, for example, have become much more assimilated than have Chinese in Malaysia or Indonesia. In the latter the social definition remains much stronger and more salient than in the former.

Basic Issues: Size and Vital Rates

Here we can indicate some of the most important population measures to include in a typical NPER, with a brief discussion of their importance for the population–environment relationship. No national review will really be typical, and it is especially important to develop measures suited to the local situation. In any event, the following measures will usually be the minimum important conditions to note in any effort to integrate population into strategies for sustainability.

Population Size and Density

A population's total size and the area in which it lives constitute some of its most important characteristics. For millennia population numbers have indicated power, and governments have typically been interested in having larger populations. But larger populations also demand more resources and can make a larger impact on their environments. Their density also indicates both intensity of impact and the extent to which they have shaped the environment to support themselves.

Population size is important not only as an indication of overall capacities for production or demands on consumption. It is perhaps more important for its indication of the infrastructural demands to provide the services and resources needed to build human capital. Large populations need more schools and health centres than do smaller populations. They need more food and water, more housing, more jobs, more utilities, and more facilities for recreation. Providing these things has become a

major responsibility of the governments that, in the same processes, are attempting to plan for sustainable development.

Density is an important ecological variable related to carrying capacity. Carrying capacity is, however, a term that has generated a great deal of confusion and controversy, especially in connection with population–environment dynamics. It is sometimes taken as a static condition, but it is obvious that it depends on technology and social organization. Wet rice can support a much higher population density than dry rice production, without drawing resources from outside its own immediate ecosystem. And the differences between wet and dry rice are clearly technological and organizational. The term is more useful for species other than the human, since non-human species tend far more to adapt to than to change their environments. By contrast, the human species produces dramatic changes in the environment. Moreover, some human populations can increase the carrying capacity for themselves by drawing on the entire global resource base, quite outside of their own immediate environment.

Thus the term carrying capacity may be useful in that it suggests limits on what an environment can support. Its imprecision and its dependence on technology and social organization, however, make it far less useful for the human species. For the most part, we avoid the concept here.

Asia has been and remains the most densely settled of our major world regions, despite large tracts of desert and sparsely settled land. Its 113 persons per square kilometre is higher than Europe's 95, and five times higher than the densities of Africa or the Americas. Europe's high density is supported by massive imports of raw materials, natural resources and energy, and exports of processed goods; that of Asia is supported far more by local production. This would imply that Asia's basic land carrying capacity is much greater than Europe's. On the other hand, it is Europe's technology and social organization that have permitted it to carry this heavy density. The Americas and Africa are distinguished by low overall densities, and they, of course, differ remarkably in the general standard of living of the populations. This, too, reflects differences in the natural environment, but even more in the technology and social organization.

Density figures can be misleading, of course, when there is a large discrepancy between overall territory and arable land. Egypt is a classical example of a country with a great discrepancy between total and arable density, but Japan and Iran can be added as well as many other countries with extensive desert areas. For many countries the relationship is roughly similar, so that a simple correlation of national measures of total and arable density tend to be rather high. For specific countries, however, the differences are striking, and it is preferable to use arable or agricultural density to get a more valid measure of the ecological condition.

Another Dimension of Density

The discussion of population density above refers to whole world regions or countries. Cities are points of exceptionally high density, but social scientists have typically found mixed results in the attempt to link density to urban problems, or to the quality of life. After all, high density can indicate affluent living in high

Table 1: 1990 Population Density and Urbanization in Major World Regions					
	Africa	Asia	Latin America	North America	Europe
Population (1990, mill.)	643	3,118	441	277	509
Density (pers/km^2)	21	113	21	13	95
% Urban	32	32	72	75	72

Source: Population – UN, 1992;[17] Urban – UN, 1994a

rise condominiums, or the cramped quarters of tenements for the very poor. The following box (p. 104) shows how density can be broken down into its component parts, each of which may indicate something very different from the others.

Mortality, Fertility and Growth Rates

These are the fundamental dynamic conditions of a population. The three rates are tied together in their 'crude' form. Crude death and birth rates are simply numbers of deaths or births per 1000 of the population, and the difference between these is the Rate of Natural Increase. They are crude and can be misleading because they are not related to the age structure of the population. Thus an older population, with larger proportion of people in the reproductive ages (15–45) and older ages (over 65), can have higher crude birth and death rates than a population that is younger, or has a larger proportion of its population in the ages under 15, even though women may actually be having more births, and people at all ages may be dying more, in the latter. Thus it is common to adjust both death and birth rates for age differences. This can be done through standardizing populations (that is, making the assumption that all have the same age structure).

17 For the population data, we have used the 1992 revision rather than the more recent 1994 revision. The latter changes what have been seen as common density patterns for the past 40 years, because of the breakup of the Soviet Union. This has produced a new and as yet unfamiliar distribution of new, and generally low density, states among Asia and Europe.

Components of Population Density

Population density can mean many different things. The following chart shows how simple density, or Population/Area, can be disaggregated into different conditions, which indicate different levels of wealth or poverty. Column 1, for example is usually described as 'crowding' when the value is high, which indicates poverty. This is also the measure where the strongest statistical associations are found with different aspects of social disorganization such as crime, drug abuse etc. High values in column 2, however, generally indicate wealth or affluence. In column 3 high values can indicate either poverty or affluence, since either wealthy high rise condominiums or crowded tenements would be included here. In column 4, we find a collective, an ecological or social condition rather than the individual or family conditions described in columns 1–3. High values here imply little open space, parks or recreation areas, and thus a kind of social or collective poverty. Low levels imply much open space, which can be described as a kind of social affluence. (Dr. Wouter Veening of the Netherlands suggested this last point.)

DENSITY IS:	1.	2.	3.	4.
$\dfrac{\text{Pop}}{\text{Area}} =$	$\dfrac{\text{Pop}}{\text{Rooms}}$ x	$\dfrac{\text{Rooms}}{\text{Household}}$ x	$\dfrac{\text{Households}}{\text{Structure}}$ x	$\dfrac{\text{Structures}}{\text{Area}}$
When the value is: HIGH =	Poverty	Affluence	Poverty or Affluence	Social Poverty (no open space)
When the value is: LOW =	Affluence	Poverty	Poverty or Affluence	Social Affluence (much open space)

Infant Mortality Rate and Total Fertility Rate

IMR
is the single most sensitive indicator of the wellbeing of a people.

A simpler and more useful procedure is to use two measures that are independent of the age structure, and are sensitive indicators of real levels of both mortality and fertility. The Infant Mortality Rate or IMR is usually considered one of the single most sensitive indicators of the health and wellbeing of a population. It is the number of deaths to infants in their first year of life, stated in proportion to 1000 live births. The range found in national societies goes from a low of 5–10 in Japan and the Scandinavian countries to highs of about 150 in parts of Africa and Southern Asia. A figure of 100 indicates that roughly 10 percent of all children die before their first birthday. This is a sensitive measure of welfare, since it indicates how well a society can care for its most vulnerable members. The IMR is also the single most powerful determinant of life expectancy at birth, which is another measure commonly used to indicate a population's physical wellbeing. An additional measure, deaths of children under five years of age, also provides a good indicator of well being, but it is far less available than the IMR. If strategies for sustainable development seek to improve and maintain the wellbeing of people and ecosystems, the IMR is perhaps the best single indicator of the portion of this aim that concerns people.

Infant deaths are relatively easy to control, given an emphasis on primary health care and a rather simple technology to prevent infectious diseases. A government concerned with popular health and wellbeing, especially of the poor and the rural population, can develop a good primary health care system, and can reduce infant mortality relatively quickly. Thus the IMR may be taken as an indication of the government's commitment to promoting wellbeing, especially of the rural poor.

An important insight can be gained by separating rural from urban estimates or measures of infant mortality. This, too, provides a good indication of government's commitment to wellbeing. It also provides an indicator of a type of government intervention that can have an especially high return. High infant mortality can be effectively addressed, and this goes a long way toward realizing the aims of a strategy for sustainability.

The total fertility rate (TFR), roughly defined as the number of children a woman will bear in her reproductive life,[18] is one of the most commonly used measures of fertility. It is adjusted for differences in age structure, so it can be compared among different populations and across time. Most 'traditional' societies, or developing economies, have shown total fertility rates of 6 to 7. All societies have shown patterns of fertility control. The theoretical biological maximum TFR is about 16, though the highest recorded levels are between 9 and 12 (Livi-Bacci, 1989). The common levels thus indicate some form of control in almost all societies. In the demographic transition, total fertility rates typically fall to 2 or below. If a society has a total fertility rate of 2.1 over time, births will come to equal deaths and the population will be stable. In other terms, fertility will have reached replacement level (TFR=2.1). Although the terms are not fully accurate, this fall of fertility has often been called the transition from natural to controlled fertility.

The demographic transition thus marks a transition from natural to controlled fertility. Total fertility rates typically fell from 6–7 to replacement level and below. Today almost all of the currently industrialized or high income countries show total fertility rates at or below replacement level.[19] Thus if they maintain those rates for 1–4 decades, their populations will begin to decline.

It is commonly acknowledged that fertility is affected by a variety of both individual and community conditions. Given the revolutionary development of the new fertility limiting technology, it is common to speak of 'demand' and 'supply' conditions affecting fertility. Demand conditions refer to the individual family demand for fertility limitation technology, and

include the desire for children, again for a variety of reasons. Thus if people desire to have no more children, we say their demand for fertility limitation is high. Demand is low if they desire more or many children, for whatever reason. Supply conditions refer to the availability or supply of the new technology, and generally include both the market and organized health and family planning programmes, both in private and government channels. We have touched above on the scientific controversy over which of the two forces exerts the greatest impact on current fertility.

Infant mortality and total fertility are closely linked together in complex ways through both demand for children and supply of fertility limitation facilities. When infant mortality is high, couples have an incentive to have more children in order to assure that some reach adulthood. At the same time, high fertility often directly causes high infant mortality. High total fertility usually implies that women bear early, frequently and late, and all three are known to increase infant mortality. Thus a vicious circle is set in motion. On the supply side, high mortality and fertility often indicate a weakness in the delivery system for the new mortality and fertility controlling technology. There is more to say about delivery systems below.

High infant mortality and total fertility are also important indicators of poverty, and of the neglect and isolation suffered by a large number of the world's people. They are thus part of that inequality in the distribution of wealth and wellbeing that *Caring for the Earth* proposes is both unstable and unsustainable, and must be eradicated. The new contraceptive technology available today makes contraceptive use one of the major determinants of both TFR and IMR.

TFR and IMR are related in complex ways through both supply and demand conditions

In low income countries: TFR and IMR together are sensitive indicators of the wellbeing of both people and ecosystems

18 More precisely, the total fertility rate is the sum of all current age-specific fertility rates. Thus it gives a cross section of how many children women of different ages are bearing. Usually a qualifying term is used: the number of children a woman will bear in her reproductive years if current age-specific birth rates remain stable.

19 In fact there are three 'industrial' countries with 1990–95 TFRs above 2.1: Iceland, 2.23; Moldova, 2.18; and New Zealand, 2.17. Two others, Ireland and Sweden, both show 2.1 (UN, 1994).

Contraceptive Use

A large industry has sprung up around the new contraceptive technology and the new national delivery systems. International and national organizations use modern probability sample surveys to obtain a great deal of information about the use of the new contraceptives (Ross and Frankenberg, 1993). These surveys typically include a vast amount of information on the personal characteristics of women and occasionally of their spouses and other family members as well. One factor these surveys help to measure is the contraceptive prevalence rate or CPR. This is the proportion of eligible couples practising contraception. Sometimes a distinction is made between use of modern methods (IUD, oral contraceptive pills, implants, injections, spermicidal creams and condoms) and traditional methods (including rhythm, withdrawal and abstinence). In addition, data are provided on women who have used and those who are currently using contraceptives. Usually when the CPR is referred to, however, it implies use of current modern methods, though failure to specify the meaning can cause confusion.

We noted above that modern family planning programmes, both public and private, are commonly used to distribute information, services and supplies of contraceptives. We also indicated that the strength of the family planning programme can be measured. The following Box (pages 108–109) provides evidence on both the impact of contraceptive use on fertility, and the impact of programme strength on contraceptive use.

Contraceptive use also has a marked impact on maternal and child health. This is now included in a broader concept of reproductive health, a term that was used at the Rio Summit in 1992, and that gained prominence at the ICPD in Cairo, September 1994 (UNFPA,

1994, 1995). The term has become controversial, because some religious groups see it as a cover for abortion, which they strongly oppose. Nonetheless, the issues are increasingly being brought into sharp focus in the world community. There is a very close relationship between contraceptive use and the health of women and children. The reasons for this are fairly well understood.

Low contraceptive use is associated with high fertility, as the Box below shows. When fertility is high, it implies that women bear early, frequently and late in their reproductive lives. There are high proportions of births before age 20, with less than two-year intervals, and after age 40. All three conditions are known to increase maternal and infant mortality. Thus increased contraceptive use raises the age of first birth, increases the interval between births, and reduces the age of last birth. In each case, the result is lower infant and maternal mortality. This can readily be seen in Figure 9, p. 108. Where contraceptive use is high, maternal mortality is low. The same is true for contraceptive use and infant mortality. We do not show that scattergram because it is almost identical to that shown here. The message is clear: where contraceptive use is low, women and infants suffer.

Contraceptive use is also closely related to the use of abortion. Where contraception is not available, abortion is used more frequently. Although it is condemned by some religious groups, and is legally restricted in many countries, abortion is still a major method for limiting fertility. Where contraceptive services are not available and abortion is permitted, as in the ex-Soviet Union and much of Eastern Europe, the effects on reproductive health are subtle, and do not show up in such measures as the maternal mortality rate. But where contraceptives are not available and abortion is restricted, the effect is usually seen in higher

Measuring the Effectiveness of a Contraceptive Delivery System

A simple ratio of rural to urban CPR or low to high education CPR provides a good general indicator of the extent to which a delivery system can penetrate the barriers of isolation that comes from rural living or from the lack of education. The following data for a few countries show the range of these measures and how they might be interpreted.

Country/Year	TFR	CPR	Ru/Urb	Lo/Hi Ed
Thailand				
1975	4.3	36%	67% (31/46)	59% (26/44)
1987	2.6	68%	97% (67/69)	85% (56/66)
Bangladesh				
1975–6	6.7	8%	37% (12/22)	21% (6/28)
1991	4.7	40%	73% (35/48)	54% (27/50)
Pakistan				
1975	7.0	5%	25% (3/12)	18% (3/22)
1984–5	7.0	8%	28% (5/18)	N/A
Mexico				
1976–7	5.0	32%	33% (13/45)	23% (13/56)
1987	3.6	53%	56% (35/59)	36% (25/70)
Kenya				
1977–78	8.1	7%	46% (6/13)	28% (4/14)
1989	6.8	27%	84% (26/31)	31% (18/35)

In these cases, as the contraceptive prevalence rate rises over time, fertility declines. Pakistan as yet shows no movement, despite announcing a policy and launching a family planning programme in 1960. Even its urban and more educated women do not use contraceptives. The delivery system scores are the lowest of all countries shown here.

In the other countries we can see an increase in the effectiveness of the delivery system over time, closely associated with both contraceptive use and fertility decline. Thailand is now a fully contracepting society and its current total fertility rate is estimated to be below replacement level. Note that rural women have as much access to contraceptives as do urban women, and even women with no education are close in access to those with more than seven years' worth. Even Bangladesh, estimated to be one of the world's poorest countries, has a delivery system that has increased in effectiveness over time, and now reaches a third of rural women (35%), and a quarter (27%) of those with no education. Mexico has made gains but has a major gap in services to rural areas and to the less educated. Kenya's overall CPR is low, but it has made major progress in the past decade in reaching the rural poor.

maternal mortality rates. It is also attested to by the medical profession, which experiences heavy demand for care from women who suffer the consequences of unsafe abortions.

A dramatic bit of evidence for this relation between contraception, abortion and maternal health comes from Romania's policy changes over the past 35 years. Like most communist countries, abortion was freely available, but contraceptive services were almost non-

Contraceptive Use, Total Fertility and Family Planning Programme Strength

The scattergrams below provide information on the impact on fertility of family planing programmes and contraceptive use. Both the major trends and the outliers are informative.

Total Fertility and Contraceptive Use

The first scattergram shows that total fertility is closely and negatively related to contraceptive use. The more people use contraceptives, the lower is the total fertility. But there are deviant cases.

The cluster of 12 countries in the lower left hand corner can be called deviant cases, in that they show a much lower fertility than would be expected by the low level of contraceptive use. All of these countries but one (Albania) are from the ex-Soviet Union. It is well known that family planning services were not well developed in the USSR, but women had free and extensive access to abortion. For these countries, abortion is the major method of fertility limitation. This is contrary to the World Plan of Action adopted at the Mexico International Conference on Population (ICP) in 1984. That stated that abortion is not a major method for family planning.

Contraceptive Use and Family Planning Programme Strength

The second scattergram shows a close positive relationship between family planning programme strength and contraceptive use. Stronger programmes produce higher contraceptive use. It is well understood that they do this by making information and services widely available, especially in rural areas and among the poor. Here, too, however, there are interesting deviant cases.

Contraceptive use and total fertility

Family planning strength and contraceptive use

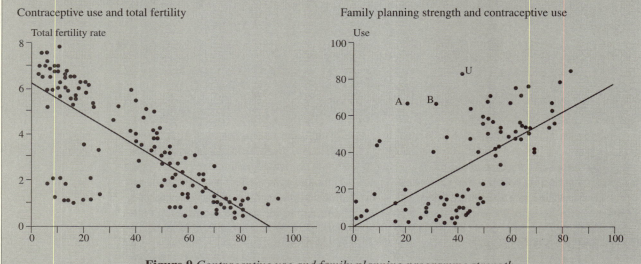

Figure 9 *Contraceptive use and family planning programme strength*

There are two sets of outliers. The three countries in the upper left (A, B, and U) are Argentina, Uruguay, and Brazil, with higher levels of contraceptive use than would be expected by the weakness of their family planing programmes. All three are considered upper middle income countries with real parity purchasing power (PPP) per capita incomes of $8,500, $6,000, and $5,200 (UNDP 1995). All are highly urbanized with well developed market systems that can constitute effective delivery systems for contraceptives even in the absence of well developed family planning programmes. They are quite different, however, in other social services, especially those relevant to reproductive health. Argentina and Uruguay have been highly urbanized for the past half century; Brazil has experienced this transition only recently, moving from 37 percent to 78 percent urban from 1950 to 1995. Argentina and Uruguay have well developed primary and reproductive health services, with over 90 percent of women receiving prenatal care and having their births attended by professionals. For Brazil the figures are just under 75 percent. Argentina and Uruguay have infant mortality rates of 24 and 20; for Brazil it is 58. Brazil's rapid transformation together with considerable economic stress, especially on the poor, and the lack of an effective family planning programme has led to the extensive use of abortion and sterilization for fertility control (Martine, 1995). One result is a high maternal mortality rate, and hospitals filled with women suffering the effects of botched abortions. Uruguay, with the highest contraceptive use rate, has only 36 maternal deaths per 100,000 live births, compared to Brazil's 200.

The other set of outliers in this scattergram are the score of countries at the middle bottom, where family planning strength is in the moderate range of 20 to 50, but contraceptive use is less than 25 per cent. All of these are among the world's poorest countries; most are in sub-Saharan Africa and in many family planning programmes are of a more recent origin. It is possible that the programme impact will increase in the near future. It is also possible that these conditions must mitigate against even relatively well developed family planning programmes.

existent. In 1966 abortion was suddenly made illegal and severe restrictions were placed on its use, with extensive monitoring by what came to be called the genealogical police, and heavy fines and prison sentences for doctors and women. The abortion-related maternal mortality rate rose from 20 deaths per 100,000 live births in 1965 to over 100 in a decade, then to above 160 in the 1980s. In 1990 abortion was made legal again, and the deaths dropped precipitously to 40 in 1991 (Stephenson, 1992).

The Delivery System

For the less developed countries as a whole, the rapid declines we have seen in fertility are very much a product of organized efforts to develop delivery systems to bring the new technology to the people. That the delivery system is not fully effective, especially for the poor, is indicated by survey data indicating that many couples want no more children, but are not able to practise fertility limitation. This is called the un-met demand for fertility limitation.

Estimates for the amount of un-met demand run for about 100 to 300 million women worldwide.

It should come as no surprise that such delivery systems usually are not as effective for the rural and poor as they are for the urban and the wealthy. These delivery systems include

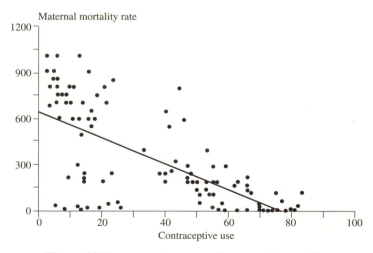

Figure 10 *Contraceptive use and maternal mortality*

market and non-market, and governmental and non-governmental organizations. The heaviest responsibility for the delivery systems, however, rests with national governments. The responsibility is both direct and indirect. Directly, governments have stated aims and aspirations of providing good health services for their populations, and especially for the rural and the poor. Governments can build effective delivery systems by making this an important goal and allocating appropriate resources for it. Indirectly, governments can improve delivery systems through educational investments, which give people greater access to all services, and by legal and economic changes that promote rather than obstruct the market as a delivery system for the new technology.

Assessing the effectiveness of a delivery system

A delivery system for any service (such as health or education) or product (from Coca-cola to contraceptives) must overcome a series of barriers that keep people isolated from the good or service. Distance, or the physical isolation of the rural areas must be overcome. This requires a good logistical system and is affected in part by the physical infrastructure (roads, trains etc) that a country has built up. In addition there is what can be called social isolation, as when illiteracy or ignorance cut people off from useful information. One can also think of economic isolation that comes

from poverty.

There is a simple and useful disaggregation that can be used to assess the effectiveness of a country's delivery system by noting its capacity to overcome the effects of physical and social isolation. National survey data on contraceptive use generally give us the proportion of women or couples using the modern contraceptive technology. These data are also often broken by rural and urban populations, and by the educational level of the women. If we assume that ruralness imposes a kind of physical isolation on rural people, then the extent to which rural people are reached by a delivery system measures the effectiveness of the system in overcoming this barrier. If we simply divide rural CPR by urban CPR, we get a sense of how effective the delivery system is in penetrating the physical isolation of the rural areas. Social isolation can be assessed by noting the impact of education on contraceptive use. Typically more educated women use contraceptives more often and more readily than less educated women.

Thus dividing the CPR of women with no education by those with a high level of education gives us a measure of the extent to which the delivery system can penetrate the isolation brought by illiteracy and lack of education. The Box on page 107 shows a series of these calculations of penetration indexes together with CPR and TFR measures of countries representative of the range of effectiveness of the new delivery systems.

Age/Sex Composition

A population's age and sex composition tells a great deal about what kind of demands the population will make both on the environment and on its own forms of social organization. Age/sex structures also tell us a great deal about what has happened recently to a population's vital rates. It is common to use population pyramids to display age/sex composition. These can be very useful for the way they enable us to visualize the impact of past conditions, to forecast futures.

The following figures illustrate how the population pyramid can tell us about both the past and the future. The first three pyramids illustrate the impact of high and low fertility when mortality is relatively low. (The crude death rates are 13.2 for Kenya, 8.6 for the US and 10.4 for Denmark.) With recent rapid declines in mortality and sustained high fertility, Kenya shows a broad base, with a large portion of the population under 15 years of age. This also provides a substantial built-in momentum for future population growth, discussed below. Even if the TFR declines rapidly, the past high fertility will leave the country with a large group of women in their reproductive ages for the next four decades. The pyramid for China in 1989 shows clearly the impact of the great famines that followed the disastrous Great Leap Forward in 1958–9. It also shows the impact of the rapid decline in fertility that came after 1972, when China adopted a very strong national fertility limitation policy.

The pyramid for Germany traces clearly the massive losses in two world wars. In the United Arab Emirates, 1985, we see reflections of the worker migration noted above, and we also can see a condition that is not uncommon in many urban areas of the developing regions: young males migrate in search of work, leaving women behind to tend the fields. The social condition may be as unstable as the figure appears.

Population pyramids can be constructed for the nation as a whole, and for any smaller unit for which there is a reasonable count of the population by age and sex. As we show in the companion volume on local communities, they can be done easily with a portable computer for a small community as well, providing the participants of the assessment with one type of picture of their current and future conditions.

Dependency Ratios

The age structure of a population makes considerable difference in what that population can do. Part of this is captured by what is called the total dependency ratio. This is the ratio of those in the 'dependent' ages (usually under 15 years and 60 or above)[20] to those in the 'active' ages, 15–60. Table 2 provides data for four countries to show how the dependency ratio moves with changes in the age structure.

For a society with high fertility and moderate or low mortality, the high proportion of young people in the population implies a high dependency ratio: many people in dependent ages to those in the active ages. This is often considered an obstacle to economic

20 These age groups are used because census reports typically give ages in five year blocks 0–4, 5–9, 10–14 and so on. The UN Population Division publishes age breakdowns of 0–4 and 5–14, for example.

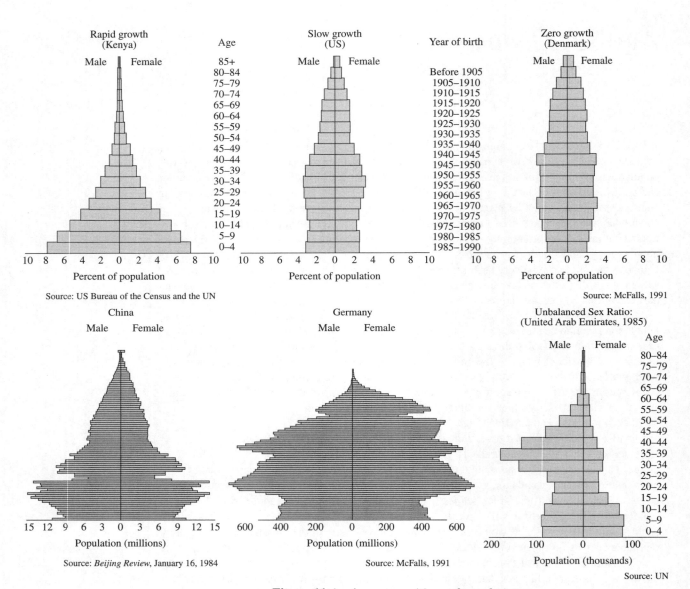

Figure 11 *Age/sex compositions of populations*

development. It implies, for example, that each person of working age must support a larger number of dependants. A low dependency ratio, on the other hand, means that each person of working age needs to support fewer dependants.

A high dependency ratio in a poor country often produces something similar to what has been called the low income trap. If a country has a large proportion of the population in the young age groups, it will have a higher cost for education. If it is poor and cannot bear that cost, many of those children will not receive an education, and will remain less productive, thus helping to keep the country poor. A wealthy country, with a low dependency ratio, on the other hand, has more resources to spend on education, and a smaller population to educate. Thus it can provide higher levels of education to its population, which will increase the productivity and allow the society to increase its wealth.

Kenya represents a common problem for the less developed regions. The rapid decline in mortality in the 1950s increased the young population to near half of the total, giving it a very high dependency ratio, which does not yet show signs of declining. Compare this with Sweden, with less than half the proportion of young people that Kenya has – and even that proportion is declining. On the other hand, Sweden has more than twice the proportion of older people that Kenya has, and that proportion is growing. This represents a future problem for all more developed regions, which experience a substantial ageing of the population.

Mauritius and South Korea show what happens to a poor, less developed region when it first experiences rapid mortality decline, but

Table 2: Age Structure and Total Dependency Ratios (TDR) for Kenya, Sweden, Mauritius and South Korea

Year	% 0–14	% 60+	TDR	Year	% 0–14	% 60+	TDR
Kenya				**Sweden**			
1950	39.8	6.3	86	1950	23.4	14.9	62
1960	45.6	6.4	108	1960	22.0	17.3	65
1970	48.1	5.9	117	1970	20.8	19.6	68
1980	49.1	5.0	118	1980	19.6	21.9	71
1990	49.1	4.5	116	1990	17.8	22.8	68
Mauritius				**South Korea**			
1950	45.2	4.7	100	1950	41.7	5.4	89
1960	46.6	4.1	103	1960	41.9	5.3	89
1970	43.8	4.2	92	1970	42.0	5.4	88
1980	37.3	4.0	70	1980	34.0	6.0	67
1990	29.8	8.0	61	1990	25.6	7.5	49

then follows this with rapid fertility decline. Both Mauritius and South Korea begin with near half the population under 15, and dependency ratios near 100. Mauritius experienced an increase in its dependency ratio to 103 in 1960, and a steady decline to the Swedish level after that. South Korea never reached the 100 mark, and its rapid fertility decline brought the dependency ratio to a low of 49 in 1990. It will undoubtedly rise in the future, as the Swedish ratio is rising, from an increase in the aged population.

Population Momentum

Population Projections and Momentum

Demographers have developed powerful statistical tools for projecting population sizes, growth rates and composition (Haub, 1987, Lee 1978). By making assumptions about future changes in fertility and mortality, one can calculate what any current population will be for some time into the future. Statistical and computer capacities make it possible to make such projections for hundreds and even thousands of years into the future. The accompanying box shows the most recent projections made by the UNDP. It should be noted that these projections typically do not take account of distinct environmental conditions, or possible changes in those conditions. Thus they tend to be environmentally uninformed.[21]

In the current projections shown in the accompanying box, we can see what many consider to be reasonable possibilities for the near future. By the middle of the next century, the world's population could reach near 12 billion, or it could begin to level off below 8 billion. Though they may appear precise, and therefore accurate, there are three important qualifications to be made about these projections.

Projections for the near future (10–20 years) can be quite accurate, since the people who will be bearing children are already born, and it is reasonable not to expect radical changes in mortality and fertility in that period. Thus the projections for the year 1990 that the UNDP

made in 1973 were found to be only 15 million people short of the actual total, or a difference of only 0.3 percent. The projections made in the 1960s were only about 2 percent off.

The further into the future we go, the more uncertain the projections become. Beyond 20 or 30 years into the future, the uncertainty increases and the potential for substantial change and errors is great. But note that despite the uncertainties, the figures are still given in very precise terms. Precision does not necessarily mean accuracy.

The smaller the area of estimate, the greater the chance for error. Although the 1973 UN estimate for the year 1990 was only 0.3 percent too low, the same estimate for Eastern and Western Africa was 9.6 percent too low, and that for Latin America was 9.3 percent too high. Africa grew more rapidly than anticipated because its mortality dropped more rapidly with no change in fertility. Latin America grew more slowly because its fertility dropped more than expected. When we deal with a small district or village the estimates can be very wrong because even small migration flows can overwhelm small communities. Although migration flows tend to be somewhat restricted across national boundaries, massive refugee flows are not unknown and may well grow more important in the future. Demographers do not have tools that can provide very good projections of population migrations.

One of the most important uses of projections is to provide a clear illustration of population momentum. This is the condition that comes from the age structure. If a society

21 There is one way in which projections do take some account of environmental conditions. It is common to build life tables from actual population conditions for different population groups, such as those for the developed countries, and those for various regions in the less developed world. To the extent that these regional differences reflect environmental differences, it can be said that environmental conditions inform population projections. The environmental conditions are only general, however, and are not specified.

moves from high to low fertility, the large numbers of girls born when fertility was high will take some time to reach child-bearing age. Thus even if women now begin to have fewer children, the larger numbers already born will mean continued rapid growth for up to 40 years in the future.

Under current conditions where mortality control can be facilitated by modern medical and public health technologies, it is relatively easy to begin rapid population growth. Growth is more difficult to slow down, however, because of the distinctive built-in momentum.

Relatively simple health interventions, such as malaria control, oral rehydration therapy, or a vaccination programme can reduce infant mortality quickly. Public health interventions such as protecting water supplies or safe disposal of human wastes can have the same impact.

When infant mortality declines without parallel declines in fertility, rapid population growth is set in motion. Under these conditions, the proportion of young people in the population grows; the population gets younger.

Increasing the number of infants has a number of different future impacts on the population. More infants today means an immediate demand for more health services. In five to ten years those new infants will need more school places and in 15 to 20 years they will need more jobs or land. Perhaps more important, in about 15–20 years the increased number of female infants will enter reproductive age, and remain there for another 30 years. Thus even if the total fertility rate declines rapidly, population can continue to grow as the large numbers of children from past high fertility mature and enter the reproductive age.

Following (Figures 10a–c) are graphic representations of the two communities, each of which began with the same population of 2,100.

The Powers and Limitations of Population Projections

Working with just a few variables – death, birth and migration assumptions – demographers can generate highly precise (though not necessarily accurate) projections of a population some years ahead. Changing assumptions can provide a variety of different scenarios (or variants in the language of demography).

The UN provides a periodic assessment of the world population prospects, which is revised every two years, and gives a wide range of population data, in five year periods, from 1950 to 2025 for all countries of the world. The most recent projections, the *1994 Revision*, extends the horizon to the year 2050. The projections for the future include three variants: low, medium and high, reflecting different assumptions about the future course of both birth and death rates. The 1994 revision of the *World Population Prospects* gives the following three estimates for the total world population in the years 2000, 2025, and 2050, for example.

	2000	2025	2050
Low variant	6.081	7.603	7.918
Medium variant	6.158	8.294	9.833
High variant	6.235	8.979	11.912

The graphs show clearly the impact on different age groups, and the delay in the impact of rapid fertility decline. The point is that the different impacts take some time to be felt.

Total Population

Total population remains similar for almost 30 years, since mortality is falling more rapidly in the community with more rapid fertility decline. Differences do not begin to appear for 25 years, but then they grow rapidly. After 50 years, however the slow decline community has almost twice (1.7 times) the number of the rapid decline community. Moreover, in 50 years the rapid decline community is only growing at less than 1 percent per year, while

An Illustration: The Future Implications of Changing Fertility

Here are two hypothetical communities, with demographic characteristics that are quite common for many developing countries. They both begin with 2,100 people, high fertility and mortality, and age/sex distributions common for many less developed countries.

Then we assume that Community 1 shows very SLOW declines in both mortality and fertility. Infant mortality declines slowly, remaining above 100 until after 2005, and life expectancy for males and females together increases only slowly from 38 to 58. The total fertility rate declines only gradually down to 5 by the year 2020. This is, in fact, a scenario close to that projected for many African countries.

We assume that Community 2 shows the kind of RAPID decline in mortality and fertility that Thailand, China, Korea or Taiwan showed after 1965. Infant mortality goes down from 127 to 37 in 25 years, and life expectancy for males and females together rose in that same period from 49 to 73. The total fertility rate declines from 7 to replacement level in 25 years (1970–95).

Total Population Showing males and females for five year periods 1970–2020

Year	Community 1 (SLOW mortality and fertility decline)			Community 2 (RAPID mortality and fertility decline)		
	Total	Males	Fem	Total	Males	Fem
1970	2100	1045	1055	2100	1045	1055
1975	2360	1177	1183	2413	1212	1219
1980	2663	1330	1333	2780	1390	1390
1985	3024	1513	1512	3127	1567	1560
1990	3461	1731	1729	3460	1735	1725
1995	3996	1998	1998	3756	1883	1872
2000	4620	2309	2311	4008	2010	1998
2005	5338	2668	2669	4256	2134	2122
2010	6184	3093	3091	4499	2255	2244
2015	7176	3592	3584	4718	2366	2353
2020	8277	4146	4131	4899	2456	2443

The figures on the next page show how these differences are worked out over time, and their implications for different age groups.

the slower decline community is still growing by almost 3 percent (2.8). At those rates the slower decline community will double in 25 years, whereas the rapid decline community will require almost 90 years to double.

Infants

The most rapid impact is seen in the 0–4 year age group. There is first a more rapid rise in the community with rapid fertility declines, because of the rapid decline in infant mortality as well. But then the numbers drop off very quickly. Within 30 years, the first generation,

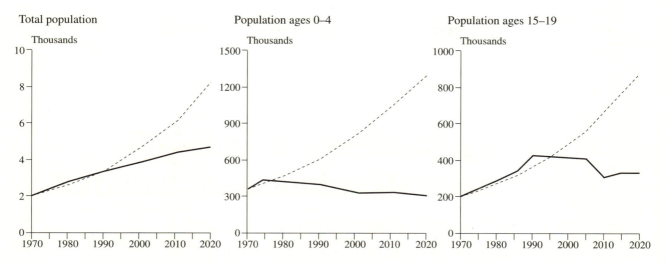

Figure 12 *Total population growth among infants and young adults, 1970–2020*

the slower declines produce more than twice as many 0–4 year-olds (829 vs. 342) as the community with rapid declines. This implies a reduction of costs for child care, or a higher allocation per child. It also shows the close relation between fertility and infant mortality. A decline in fertility has a favourable impact on infant mortality. This is just the beginning, however. At the end of this 50-year period, slow fertility decline leaves the community with almost 2 percent per year growth in these young ages, while the rapid fertility decline means that infants are declining by about 1 percent per year.

Young Adults

A major difference among the young adults, 15–19 years, is not seen for more than 30 years. Thus one cannot count on rapid fertility reductions to reduce the demand for land or jobs for at least a generation. By 50 years, however, the differences are large and growing.

Then the slow decline community must find jobs or land for 875 new young people, as against only 344 in the rapid decline community. Moreover, at the end of 50 years this young adult population is growing at the rate of 2.5 percent per year where fertility declines have been slow, and is increasing by only 0.2 percent where fertility decline has been rapid. And that growth is only an echo of earlier growth, which will decline shortly. One cannot look for early relief in the slow fertility decline community, which will continue to show high growth rates of the young adult population for some time.

The projections shown above were done using a simple and readily available computer program, DEMPROJ.

DEMPROJ can be obtained at little or no cost from the Futures Group, 1050 17th Street, Washington, DC, USA 20036: Fax 1 202 775 9694.

Migration and Environment

22 This is perhaps the most extreme estimate of potential migration disasters to be found. It is criticized as being too alarmist in that it would occur only under truly massive upheavals. It is useful, however, to illustrate an important implication of rapid population growth. The very large size dwarfs any past experience we have.

The history of the human species is a history of migration. From small centres in Eastern Africa, humans have migrated to all known places on the planet. Foods and pathogens have moved with people to create a global ecosystem, tied together by human activities. Massive migrations have affected all known land masses and all indigenous peoples, who were themselves prior migrants to sometimes empty lands. In all of these past cases, environmental conditions undoubtedly caused the migration streams, and were subsequently greatly altered by those streams.

Cause:

Environmental degradation or overcrowding pushed people to other, more hospitable environments. There have probably always been what we now call environmental refugees.

Impact:

Deforestation and expanding agriculture in the Americas, expanding agriculture and plantations in Central America and Southeast Asia, and expanding oil production in the Middle East are only some of the more recent and dramatic examples of the environmental impact of migration.

In one respect, the current migration–environment relationship is merely an extension of the past common behaviour pattern of the human species. There are, however, new conditions in the current situation, which only exacerbate the relation-

Bangkok is subsiding 13 centimetres per year due to ground water withdrawals. Combined sea level rise and land subsidence in Egypt and Bangladesh could produce 48 million environmental refugees.

Jacobson, 1988

ship between migration and the environment. One is the magnitude of the movements, which is directly related to the magnitude of the world's population. Our larger population inevitably implies movements of people that will dwarf even the largest of past movements.

The number of environmental refugees is growing steadily, and may reach one billion (nearly a sixth of the world's population) by the end of the century. (RPG, 1992).[22]

In the past decade an estimated 154 million hectares of tropical forests were cleared for other use. Since 1945 an estimated 2 billion hectares of rangelands have been degraded, 700 million hectares of this by livestock overgrazing. (WRI, 1994)

In addition, the human species has created a new chemically based industry, which produces toxic accidents, and environmental wars that degrade the environment and produce hundreds of thousands of environmental refugees. The names of Bhopal and Chernobyl, the herbicides used in Vietnam and El Salvador, and the poisons used in Afghanistan and against the Kurds are only the more well known examples of this new aspect of environmental degradation (Jacobson, 1988).

Large scale development activities, such as dams, roads and ports create more thousands of environmental refugees. The prospects of global warming and sea level rise from world wide development offers the prospects of unprecedented millions of environmental refugees.

Finally, because the numbers and densities are so large today, disruptions from natural disasters and political strife can produce massive numbers of refugees that overwhelm the environment of a receiving area. More than a million Cambodians have been killed. A million Rwandans on the move has caused an environmental nightmare. As many as 48 million people were recently uprooted by floods in Bangladesh. These anecdotes only begin to illustrate the massive impacts that occur when numbers are large and densities are high.

What is to be Done?

Of all population problems, those of migration appear the most intractable. The data are very defective, so that we seldom know precisely who is moving where. Moreover, the movements today can be so massive as to overwhelm any planning. And there are major differences in national migration policies throughout the world. Migration streams are very difficult to start when people are not willing to move, and very difficult to stop when they do wish to move.

Nonetheless, there are some practical ways of addressing the issue of migration and environmental planning. To make some concrete suggestions, we must deal with three different types of migration streams: international, domestic, and those connected especially with the distinctive problem of urbanization.

International Migration

Out-migration and remittances

National planning for international migration and sustainable development confronts two conflicting interests, streams and problems: those dealing with out-migration and with in-migration. Many countries in the low income world send substantial numbers of migrants abroad for work, which results in important income flows back to the country. This can represent substantial sums, especially for families in poor countries. It can also represent a large proportion of foreign exports, thus making it of interest to governments. The roughly $1 billion of remittances of Korean workers abroad represents only 2 percent of exports, but the $600 million of Yemeni remittances represents 150 percent of that country's exports (ILO, 1992). Governments may also wish to keep these labour flows and their remittances confidential, so they will not enter easily into the planning process. At the same time, these flows may not have much direct impact on environmental or population issues in the sending country;[23] thus they may be safely left out of the process if the data are deficient or if governments prefer secrecy about them.

Immigration

Where in-migration is concerned, the population and environmental problems are more serious, and more intractable. A country may be subject to small immigration flows that can destabilize fragile environments, and lead

> The World Bank estimates that remittances from migrant workers amounted to $65 billion in 1989, second only to crude oil in international trade.
>
> Source: ILO 1992

23 Virginia Abernathy (1993) has argued that such out-migration does indeed have an impact on population dynamics, and there is some evidence to support this. For villages and families receiving remittances for their migrants, high fertility may represent a very rational response to poverty. On the other hand, it is doubtful if the size of the migrant stream has much impact on a nation's fertility level, at least for most countries of the world.

> ## *Lessons from Natural Hazard Studies*
>
> Vulnerability reduction measures can be taken and are cost effective, either as stand-alone projects or as components of overall sector development programmes.
>
> Sectoral studies reveal previously unrecognized linkages between development and natural disasters.
>
> A sector may have to select between competing objectives to arrive at an acceptable vulnerability reduction strategy.
>
> Source: OAS, 1990

24 This is the case in parts of the Sahel, for example. See Agbo et al, 1993.

25 Theoretically governments have the right to control their borders, and many go to consider-able expense to exercise that right. Few are totally successful, and fewer still would be capable of stopping the kind of mass migration of Rwandans that Zaire experienced, or even the smaller numbers of Albanians that have given the Italian government such difficulty. The US's experience in migration control on its southern borders is only one illustra-tion of how difficult it is in fact to control borders.

to ethnic conflicts with indigenous peoples.[24] It may also be subject to massive immigration streams from either political strife or environ-mental disasters, or both. Famines, for example, can be caused by natural disasters such as drought, but they are far more often caused by political conditions. Floods, earthquakes and other natural disasters often force large masses into temporary movement. People return when the environment returns to its prior state.

For planning, a national government can map its borders, estimating the pressure for small scale, gradual movements, and the potential for large scale refugee movements.[25] Overlaying this with a map of border environ-mental conditions, noting especially vulnerable habitats or fragile environments, will produce data from which a series of future scenarios might be simulated, with suggested policy options for each. That is, where these interna-tional refugee flows are concerned there is probably little a government can do other than to attempt to anticipate them, so that when and if they do occur, the government is not caught completely unprepared.

There are also useful tools for disaster planning, which can help a government to map potential problems and develop future scenarios. GIS, remote sensing, and special mapping techniques have been developed to assess potentials for natural hazards. Related techniques of natural hazard, vulnerability and risk assessment have been developed and can readily be integrated into overall strategies for sustainable development (OAS, 1990). Studies from the development and use of such tools offer important lessons. Where natural disasters cannot be avoided or precluded, planning and foresight can reduce the vulnerability of both the population and the environment to such disasters. In other cases, development itself may cause natural disasters, such as flooding, and

recognition of this potential is essential for more effective planning for sustainable development.

Domestic Migration

This is another area where the data are weak and the problems seemingly intractable. It is also an area where pressures from population growth appear to be especially pronounced. Population growth, crowding, and land exhaus-tion all contribute forces for internal migration. Local populations may feel these forces as powerful intrusions, but national governments typically have very little real data with which to work. More important, few governments have succeeded in stopping or promoting migration flows. The practice of proclaiming areas to be protected and forcing out local inhabitants, or forbidding the entrance of settlers, is notorious for its failures. Squatters move in, pushed by poverty or environmental degradation, and pulled by opportunities in less crowded environments.

For parks and protected areas, there has been a movement in both theory and good practice away from fences and guns to exclude people, and toward strategies like joint management that include local people and give them a stake in the protected area (Barton et al, 1996). These strategies for dealing with local migration offer much promise.

A useful planning tool for internal migration is to map what can be called popula-tion–environment hot spots. While these must be defined for a specific location, they would typically include vulnerable habitats, especially those relatively empty lands that lie close to major population concentrations, and could attract settlers or squatters. Land adjacent to new road developments will almost always attract new settlers, as in the Amazon basin and in areas of Southeast Asian logging (Grainger, 1993).

Addressing undesired internal migration streams is best done by promoting human and environmental welfare at the source of the stream, the places that are generating the outward movement.

Educational and health services, the physical infrastructure for improved life quality, and institutional arrangements such as employment creating investments and tax incentives, can all be used to affect migration flows. While these may appear costly and with little immediate pay off, careful cost effectiveness assessments often indicate that it is far cheaper to deal with migration streams early and at the origin, than attempting to deal with them at the receiving point after the flows have become large.

Urbanization

26 This even understates the degree of urbanization, as urban lifestyles and behaviour patterns spread to what are still defined demographically as rural areas. Sri Lanka and Thailand, for example, are each less than 30 percent urban, but their rural populations have adopted reproductive and consumption patterns almost identical to their urban neighbours.

27 One reader remarked accurately, however, that the pressure on the Amazon would have been far greater without the rapid urbanization Brazil has experienced.

Urbanization is one of the most distinctive features of our modern societies. Not only has the world become an urban industrialized society, it continues to become more and more urbanized. The process began in Europe and North America in the last century and is now spreading rapidly to the rest of the world as an apparently inexorable force. At present approximately 45 percent of the world's population lives in urban areas,[26] and projections are that by 2025 the proportion will have risen to over 60 percent. It is not expected to stop rising until some 70–80 percent of the world's population lives in urban areas.

Regional distributions show considerable variation. The more developed regions are now 70–80 percent urbanized and have stabilized at that level. The less developed regions are now about 35 percent urbanized and their urban populations are growing at a pace greater than their rate of population growth (UN, 1996). By 2025, they are expected to be just under 60 percent urbanized. There is even greater variation among the less developed regions themselves. South America is now as urbanized (72 percent) as North America or Europe. Africa and Asia remain the least urbanized, but they, too, are moving rapidly toward full urbanization. Africa and Asia are expected to be 54 percent urbanized by 2025.

The course of rural populations over this period shows considerable variation and may have some implications for environmental planning. For example, if urbanization implies population decline in the countryside, this may offer opportunities to extend protected areas to

support biodiversity and wildlife (McNeely and Ness, 1995). The European rural population was cut from 262 million to 192 million from 1950 to 1995. The Latin American rural population reached a peak in 1985 at 126 million, and has been declining since. The North American rural population is estimated to show continued growth to 1995, when it will be almost 69 million, and then will gradually decline.

The aggregate figures can be misleading, of course, and may not tell us much about real population pressures on rural areas. For example, Brazil's rural population began to decline in absolute numbers in 1970, 15 years before the Latin American total, but that has not eliminated the pressure of poor farmers moving into the Amazon forests to convert them to agricultural lands.[27]

The point to be derived from this is the general one taken throughout this volume. Population–environment dynamics are location specific. They should be studied in specific locations, and policies or interventions should be specifically tailored to local conditions.

But there is another important lesson as well, and one that is often not intuitively acceptable. Urbanization may well be a major force for reducing the human degrading impact on the environment. Urbanization is often decried today for its assault on the environment, and for the highly visible problems of poverty and decay that we see daily. The shanties clinging to hillsides in Brazilian favelas, and shacks that surround Manila's 'Smoky Mountain', and the heavy yellow sickening air of some Chinese cities are pictures called up when we see mounting urban disasters throughout the world. Nor are these problems of the less developed regions alone. Paris and Rome rival Bangkok or Mexico City for traffic gridlock and air pollution. Inner city Detroit rivals Calcutta's slums for degraded human environment, and may also be far more dangerous to its inhabitants as well.

Serious as urban problems are, it seems inevitable that urbanization will continue until the entire world is 78–80 percent urbanized. More than that, however, it is quite likely that this may have a beneficial environmental impact. Urban living may well be able to provide a higher quality of life with less environmental degradation than would be possible in rural living. Indeed, the argument is now being made that urban living provides a more efficient form of human settlement than rural living, and that environmental degradation will be reduced by further urbanization (Martine, 1995). There are other potential benefits for the environment as well.

As people create an urban environment, they often work to make that environment pleasant. One result is that cities also contain green areas. In addition, through public parks, zoos and gardens, and through private commercial and household activities, cities are rich in biodiversity and they are green and photosynthesizing. Often species richness in a city is considerably greater than in its surrounding countryside (McNeely, 1995). Finally, cities have always been attractive as centres of culture, creativity and innovation. When cities are well managed, they can be environmentally friendly and integral parts of strategies for sustainability, as Curitiba in Brazil has shown so well (Rabinonivitch, 1995).

Finally, the experience of the more developed countries shows that urbanization is ultimately associated with an absolute reduction in the population living in rural areas (McNeely and Ness, 1996). Rural populations have been declining since 1950 in Europe and Japan. The European rural population was cut from 262 million to 192 million from 1950 to 1995. In the same period Japan's rural population declined from 42 to 28 million. In many areas of Europe and Japan we find deserted villages alongside the high urbanization. Even in Brazil, the massive urban transformation meant a decline in the rural population starting as early as 1975. Of course, whether this reduction of the rural population will mean greater or less environmental degradation depends on a range of other conditions, including global economic forces and national and international policies. Nonetheless, urbanization will also imply some reduced pressure of the human population on large tracts of rural areas.

Thus rather than decrying urbanization and trying to halt or reverse the trends, governments would do better to examine the character of urban systems and work to make them more efficient and more environmentally friendly. To do this, it is useful to think of urban systems as metabolic processes. Before we go into this in detail, however, we should consider a problem of measurement. How big are cities? How do we define their boundaries?

Overbounded and Underbounded Cities

Urban areas are usually defined by administrative boundaries – the number or proportion of people living in administrative areas (towns, municipalities, counties or districts etc.) that are defined as urban. These administrative areas may, however, include a substantial amount of forest, pasture or agricultural land. In this case we can say that the city is overbounded, or the administrative area extends beyond the actual built up urban area.

> ### *Satellite imagery can provide information on the amount of land transformed into urban built-up areas, showing rates of change over time.*
>
> Satellite images of Bangkok taken in 1985 and 1989 were computer-enhanced to show change of land from rural to urban uses. The changed land appears in red, showing extensive ribbon development, or new urban land use, radiating outward from the centre along the road network. It also shows that 3024 hectares or about 0.3 percent of the area covered by the image was transformed from agricultural to urban land in that five year period.

In other cases, urban structures, land forms and life styles extend beyond the administrative boundaries into areas technically defined as rural or non-urban. In this case, cities are said to be underbounded, or to spill over their administrative boundaries.

Land Use or Administrative Area

An alternative to administrative areas in defining urban areas is land use. Urban areas imply densely packed buildings, roads and public structures, or a 'built-up' area. Maps can easily show built-up areas, and can also show where administrative boundaries lie in respect to the physical boundaries of built-up areas. With electronic remote sensing it is becoming very easy to obtain images that can show built up areas and their changes over time. It is also easy to superimpose (digitize) on these images the administrative boundaries of the area. This makes it both possible, and increasingly inexpensive, to obtain computer enhanced images that show the expansion of urban built-up areas over time, indicating what type of land has been converted into urban land. It is also possible to obtain precise measures of the amount of land under different land use classi-fications, and thus to know the rate at which

land is being transformed into urban built-up areas. This is another example of how satellite imagery can be an effective tool in planning for sustainable development.

Urban Metabolism

Cities can be conceived of as living systems, driven by basic metabolic processes. (The following chart provides one representation of this metaphor.) Although it should not be pushed too far or taken too literally, this metaphor has some important advantages. As living systems, cities take in people, resources and energy, and transform these into a distinc-tive quality of life (QOL). Like any metabolic process, the outcome is life itself. Like other living systems, cities produce material and symbolic goods, and they emit wastes. But all of these can vary among cities and in any city over time, showing higher or lower levels of the quality of life itself. Some cities are more efficient or more healthy than others; they use less energy to produce a higher quality of life while emitting less waste. Many cities in Japan and the western, more developed regions appear to have improved their efficiency or metabolic health over the past two decades. Cities in Eastern Europe, Russia and many less developed regions may have moved in the opposite direction.

If we can understand what kinds of technology and social organization affect urban metabolism, we should be able to design systems that are more efficient, and produce a higher quality of life with less environmental damage. A NPER should include assessments of the quality of life of individual cities and studies to determine what kinds of policies, leadership or technology account for difference in the quality of life. Quality of life indicators should include:

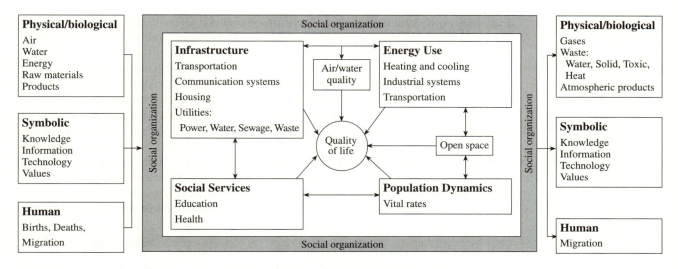

Figure 13 *Urban Metabolism Flow Chart*

- Infant and child mortality rates, by different income groups and by gender;
- safe water availability, by income groups or neighbourhoods;
- sewage and solid waste disposal, by income groups or neighbourhoods;
- educational enrolment rates, especially distinguishing males and females;
- safety, or indicators of violence against persons and property; and
- air quality.

These conditions refer to some of the most fundamental bases of the quality of life. They reflect the physical and institutional environment that sustains life and builds human capital. Until these basic conditions are in place, few other QOL measures will be important. When they are met, however, other conditions gain greater importance. Thus opportunities for employment, housing, recreation and the development of human talents can be included in QOL measures, after the basic conditions are met.

Urban impacts on the environment extend considerably beyond the transformation of land. Cities and towns demand resources from outside, and empty their wastes beyond their own boundaries. The high concentrations of people in these restricted spaces greatly amplifies all human impacts. These resource uses and waste dispositions should be assessed for the nation as a whole, and for its administrative areas, with both current estimates and future projections. In addition, the following conditions indicate important linkages between urbanization and the environment and should be an integral part of the basic NPER:

- Water requirements, sources, use, destination, and quality.
- Human waste amounts and management, including solid waste generation and disposition.
- Energy requirements, types and sources of energy.

Measuring the Quality of Life

It is easy to understand that the quality of life varies. Some individuals, communities or nations eat better, live longer, suffer less illness, experience greater personal safety, and have more opportunities to develop their own talents than others. The measurement of QOL, however, has for some time presented difficult problems. The national income account has provided some help, and is especially useful since it collects into a single measure a wide diversity of individual and group actions. That it correlates so highly with other welfare measures, such as life expectancy, makes it a candidate for assessing quality of life.

It has been understood for some time, however, that per capita GNP or GDP is deficient as a measure of life quality, and there have been many searches for a better alternative. The Physical Quality of Life Index (PQLI) was developed some years ago by Morris David Morris (1977). This combined infant mortality, with life expectancy at age one, with adult literacy into a single index from 0 to 100. Although this was useful, especially at distinguishing different welfare accomplishments at the middle levels of per capita GNP, it was found to be highly dependent on the IMR, and it has not gained much currency.

More recently, the UNDP has generated a new measure, the Human Development Index (HDI) (UNDP, 1993). This combines measures of wealth, health and education. Wealth is

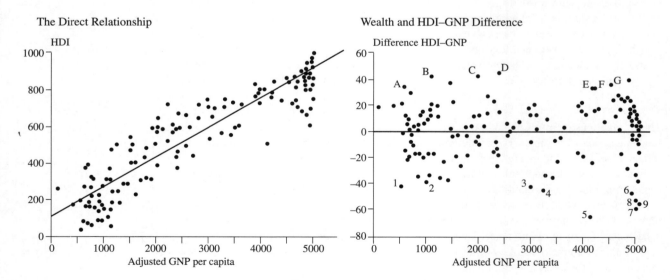

Figure 14 *Wealth and human development I and II*

indicated by a per capita GDP measure that is adjusted for international comparisons in purchasing power. Health is measured by life expectancy, and education has two indicators: literacy and mean years of schooling. All three measures are combined into an HDI that is expressed from 0 to 1,000 in three decimal places. The calculation has been made for 173 countries. In addition, all countries have been ranked from 1 (high) to 173 (low).

The HDI represents a good step forward in assessing QOL, but it is better for making large scale comparisons among nations than for making the more refined distinctions between sub-national areas and between different social groups that a national government would wish to make. As a practical matter, the HDI is itself highly correlated with life expectancy (r= +0.96).[28] Thus it is much easier for a national government to use life expectancy as a surrogate measure for QOL.

The HDI analysis makes another useful suggestion, especially for countries at different levels of economic development (UNDP, 1993, page 112). At low levels of development, more basic measures of life expectancy are the most useful. At higher levels of development, however, the general measure of life expectancy rises for all people and groups, so that it becomes less useful as a measure of life quality. Thus the simpler and more readily available measures of life expectancy or infant mortality provide good measures of QOL for less developed regions. For higher levels of economic development, however, measures such as under-five mortality and maternal mortality will provide more sensitive measures.

There is one additional significant condition that the HDI captures. The UNDP has examined the difference in rankings between HDI and per capita GDP. The following scattergrams reveal something interesting about this measure. First note that in Figure 12a, the HDI is closely related to wealth, or the adjusted GDP per capita. But then we examine the differences in GDP and HDI ranks. The difference can be said to reflect a government or society's capacity to turn wealth into welfare or human development. Figure 12b shows this capacity is not related to wealth. Countries doing very well on this measure include A: Tanzania, B: Vietnam, C: China, D: Sri Lanka, E: Poland, F: Colombia, and G: Chile. Those especially weak on this measure include 1: Guinea, 2: Djibouti, 3: Algeria, 4: Iran, 5: Gabon, 6: Lybia, 7: Saudia Arabia, 8: United Arab Emirates, and 9: Oman. At all levels of wealth some countries do better than others in turning their wealth into welfare.

28 Where 'r' refers to the Pearsonian Correlation Coefficient

Socially Defined Groups: Gender, Ethnicity and Indigenous Peoples

Gender

29 This is not by any means to deny the biological differences between men and women. But that biological difference is constant among all societies, and societies differ greatly in the social definitions they give to these biologically differentiated individuals.

All societies differentiate between males and females, defining certain roles, work, rights and obligations for each. These can vary immensely from society to society (Whyte, 1978). Moreover, what one society defines as appropriate only for men (e.g. work, physical behaviour, or even certain forms of adornment), another might define as appropriate only for women. In some respects these differences may appear arbitrary, but they have probably emerged gradually over time as rational responses to basic ecological conditions. Most of these definitions are rooted in history and are seen by people not simply as rational, but as the right way, often the only right way, to do things.

In almost all societies men have more power, more freedom and more control over resources than women. Nowhere is there gender equality; societies differ only by the amount of inequality women experience. Everywhere these differences tend to be deeply

held and to be supported by the religious beliefs and organizations.

No matter how deeply held these beliefs are, and whatever reason is given for them, they are still fundamentally social definitions.[29] This implies that they can change; anything that is socially defined can be socially redefined. We have recently seen, for example, women gain the right to vote, the right to an education, the right to enter previously closed professions, and even the right to wear clothes previously appropriate only to men. Moreover, world organizations, such as the UN and some religious and political groups have called for new universal social definitions giving women greater equality. The UN ICPD in Cairo in September 1994 articulated these aims with great vigour (UNFPA, 1994). These groups are supporting major changes in the social definitions of gender roles.

Although such definitions can change, they often do not do so easily. When they are deeply embedded, as they are when supported by religious precepts, they tend to be seen as right. Proposals to change them often meet with strong rejection, and sometimes even with violence. It is still not well understood, however, how and why social definitions change, but they do change, and sometimes very rapidly. What may appear unthinkable one day, can sometimes be fully accepted in a very short while.

The relation between gender roles and population–environment relationships is clear, but controversial. The new contraceptive technology, which gives women new and

---

Changing Gender Roles

In the early 20th century, when women began to enter office work, opponents argued that they could not stand up to long hours over the typewriter. Today most of the world's secretaries are women.

Recently some medical doctors have been known to argue that women cannot be good surgeons, because they cannot stand up to long hours over the surgery table. Female surgeons are well established and growing in number.

Anything socially defined can be socially redefined.

powerful controls over their fertility, represents a revolutionary change in gender roles. There is ample evidence of the controversial nature of this change in the religious resistance to modern family planning programmes, and to population planning more generally. It is clear that fertility declines are very closely associated with female status, autonomy, and freedom. Where women are more educated, are accorded greater status and autonomy, and where they can move about freely, fertility tends to come down most rapidly (UNFPA, 1994).

On the environment side, it has also been noted that women are major caretakers of the earth (IUCN, 1993). In many societies they produce most of the food and care for the land, despite the fact that they usually do not have legal control over the land or access to such things as credit. In the 1970s and 80s, this led to increasing attention to Women in Development (WID) programmes, calling for greater assistance to women in both development and environmental protection. More recently the attention has shifted slightly to what is called Gender and Development (GAD), in which attention is paid to the roles of men and women, and the ways in which those roles together advance or inhibit attempts to promote sustainable development (Feldstein and Poats, 1989; IUCN, 1991; Mosher 1993; UNFPA, 1994; WRI, 1994).

Thus an NPER must take account of the relative positions of men and women, and especially to their rights to control resources, and to their access to social services. This implies an examination of legal provisions, and of the actual position women occupy in the social structure. Most indicators of health and welfare, for example, can be disaggregated by gender to provide objective indicators of access to natural and social resources. Such things as literacy and school enrolment, infant and child mortality, life expectancy and access to primary

---

Mapping Ethnicity

Mapping the distribution, size, growth rates, land use and economic activity of ethnic groups can provide valuable information for an NPER.

---

health care can be shown for both males and females, and provide good evidence of the implication of the gender role within social definitions.

Ethnicity

Much of what is said about gender can be said about ethnicity. These are social definitions that can often seem arbitrary. They are, however, rooted in history and often reflect different ecological positions and the struggle over resources. They are universal, in that all groups tend to define themselves as different from, and usually better than, others. Like gender definitions, they are deeply embedded in the social conscience, held with great passion, and often supported by religious formulations. They can also change quickly, for reasons that are not always fully understood.[30]

They are different from gender definitions, however, in that they tend to be applied to groups, often those living in distinct territories or performing distinctive tasks. They also seem to be different, especially in our present context, in the violence that surrounds the definitions.

The impact of ethnic definitions is of great relevance for population–environment dynamics. Wherever there are mixed ethnic populations, for example, population policies tend to be delicate and often conflictual. In Malaysia, Fiji, Sri Lanka, India, China, the US, and much of Africa, public discussions of population policy often run into the fear of one group that it will become a minority, or will

30 German Jews, for example, were said to have been some of the most integrated in Europe, until the Nazi leadership changed the social definition, or at least developed a radical policy based in part on past partial definitions. Hawaiian Japanese at the end of 1941 turned their loyalty from the Emperor of Japan to US and went on to man some of the most decorated combat units in the World War II. Bosnian Serbs, Croats and Muslims lived in harmony, intermarried and were scarcely distinguishable from one another until the break-up of Yugoslavia brought a violent brand of nationalist leader to the political arena. These examples of rapid change of ethnic social definitions, or at least of behaviour patterns reflecting social definitions, could be multiplied manifold.

disappear, if public fertility limitation policies are adopted. Fertility and mortality rates are often found to vary significantly by ethnic group, indicating both public discrimination in the distribution of social services, and different internal group cultures.

On the environmental side, ethnic groups often concentrate in, and sometimes control, specific pieces of the environment. Sometimes, their cultures and practices lead them either to specific types of environmental protection, or degradation. Environmental policies, such as those for protected areas, or for wildlife protection often impinge upon specific ethnic groups, sometimes strongly discriminating against the less powerful or less well organized. But such policies can provoke violent upheavals, as well as being based in a more routine violence themselves. The uprisings of Chiapas Indians against the state are only a recent manifestation of a more general phenomenon.

An NPER should include the location, size, and dynamics of distinctive ethnic groups in the society. Mapping location and land use will provide the first approximation of the environmental implications of ethnicity. As with gender, all of the health and welfare measures can be disaggregated by ethnic group to show the population implications of ethnicity. Merging geographic distributions with life quality or resource control data can provide useful pictures of the importance of ethnicity for population–environment relationships, and thus for the issues surrounding sustainable development.

Indigenous Peoples

A special category of ethnicity has emerged recently in the recognition of the distinctive position, and rights, of what are called indigenous (sometimes tribal) peoples. There is a rising awareness that certain ethnic groups,

with long tenure in a territory, have often developed special ways of using resources that are sustainable over time, and may offer important lessons for the generation of strategies for sustainable development. There is also increasing recognition that such groups should be protected by the Universal Declaration of Human Rights, and permitted to maintain their cultures and life styles in the face of incursions of other populations and the need for economic development (IUCN, 1984 and 1996; IWGIA, 1992; UNGA, 1994).

For much of past history, development and the expansion of urban industrial society has simply walked over indigenous groups, pushing them out of the way, expropriating their land and resources, and sometimes massacring them. That past reaction may be coming to an end, as those indigenous groups themselves gain access to modern legal instruments and use various forms of social movements, including violent rebellions, to promote their rights.

The relevance for population is similar to that for any ethnic group. There may be differences in fertility and mortality that can be traced either to discrimination or to different cultural practices, and have direct implications for population numbers and growth. The environmental implications are somewhat more complex. They may involve important conflicts over access to resources, and they may also involve forms of knowledge that are highly relevant for issues of sustainability.

For the most part, these are location-specific issues, and call for mapping on a national or sub-national scale, similar to that for ethnic distributions. But there is an international or global dimension as well. Wildlife products of indigenous peoples have been rendered economically of lesser value because of international treaties banning certain trade in wildlife. While these legal restrictions are designed to protect what are perceived to be vulnerable species, they also conflict with the rising recognition of the indigenous peoples rights to maintain their own cultures. Sorting out this conflict will require international debate and resolution, and may be beyond the scope of work of an individual nation.

Environmental Indicators

A National Population–Environment Assessment Report will necessarily include environmental indicators alongside the population indicators discussed above. At this time there are scores of such indicators from which to choose, and no broad consensus on either what is most important or how data should be aggregated to create appropriate indicators. There is, however, a rapidly growing industry of environmental accounting that can provide useful suggestions. The UN Statistical Office is developing a new System for Environmental and Economic Accounting (SEEA), which can be found in its latest handbook (UNSO, 1993). The Organization for Economic Cooperation and Development (OECD) has been collecting and publishing handbooks of environmental indicators since 1991 (OECD, 1994). At this writing the World Resources Institute (WRI) has produced a brief but highly useful discussion of environmental indicators, which is summarized here for this discussion (Hammond et al, 1995).

First, when thinking of environmental indicators that are useful for policy, it is common to conceive of a matrix of environmental indicators that distinguish a series of issues and, for each, suggest indicators of three different conditions: pressure, state, and response. For example, for an issue like climate change, the pressure consists of greenhouse gas emissions. The state is the concentration of a gas in the atmosphere, and the response might be to plan and work toward greater energy efficiency, a reduction of emissions, or forest conservation and reforestation measures that increase the absorption of CO_2. The following figure shows one type of such a matrix based on the work of OECD and UNEP and developed by the WRI. From this matrix, WRI suggests four types of aggregate indicators: pollution, resource depletion, ecosystem risk and human welfare. A brief discussion of each of these follows.

Pollution

Human activities treat the environment as a sink for its wastes. Air, earth and water absorb human waste products. In some cases these wastes can be turned into reusable resources for life systems. Human body wastes are biodegradable and form resources for other species. In other cases wastes cannot be broken down and remain in toxic form, sometimes for centuries. The worst of these are probably nuclear wastes, whose radioactivity can remain deadly for thousands of years. In one exercise by the Netherlands (Adriaanse, 1993), five different types of pollution have been examined. They have been measured in physical terms, and tracked over time, with an indication of the target levels of reduced emission. A brief discussion of each can illustrate how policy relevant indicators can be developed. It also shows the problems that must be solved to provide useful indicators. One of the common issue is that of sustainability. What level of pollution is sustainable? This depends in large part on the natural capacities of the earth life system to turn those pollutants into useful or benign resources.

Issues	Pressure	State	Response
I. Source indicators			
1. Agriculture	Value added/gross input	Cropland as % of wealth	Rural/urban terms of trade
a. Land quality	Human induced soil degradation	Climatic classes and soil constraints	
b. Other			
2. Forest	Land use changes, inputs for EDP	Area, volumes, distribution, value of forest	In/output ratio, main users, recycling rates
3. Marine resources	Contaminants, demand for fish as food	Stock of marine species	% coverage of international protocols/conventions
4. Water	Intensity of use	Accessibility to population (weighted % of total)	Water efficiency measures
5. Subsoil Assets	Extraction rate(s)	Subsoil assets % wealth	Material balances/NNP
a. Fossil fuels	Extraction rate(s)	Proven reserves	Reverse energy subsidies
b. Metals and minerals	Extraction rate(s)	Proven reserves	In/output ratio, main users, recycling rates
II. Sink or Pollution indicators			
1. Climate change			
a. Greenhouse gases	Emissions of CO_2	Atmospheric concentration of greenhouse gases	Energy efficiency of NNP
b. Stratospheric ozone	Apparent consumption of CFCs	Atmospheric concentration of CFCs	% coverage of international protocols/conventions
2. Acidification	Emissions of SO_x, NO_x	Concentration of pH, SO_x, NO_x, in precipitation	Expenditures on pollution abatement
3. Eutrophication	Use of phosphates (P), Nitrates (N)	Biological Oxygen Demand P, N in rivers	% population waste treatment
4. Toxification	Generation of hazardous waste/load	Concentration of lead, cadmium, etc in rivers	% petrol unleaded
III. Life Support indicators			
1. Biodiversity	Land use changes	Habitat/NR	Protected areas as % threatened
2. Oceans	Threatened, extinct species % total		
3. Special lands (eg wetland)			
IV. Human Impact indicators			
1. Health	Burden of disease (DALYs/person)	Life expectancy at birth	% NNP spent on health vaccination
a. Water quality		Dissolved oxygen, faecal coliform	Access to safe water
b. Air quality	Energy demand	Concentration of particulates SO_2, etc	
c. Occupational exposures, etc			
2. Food security and quality			
3. Housing/urban	Population density (persons/km^2)		% NNP spent on housing
4. Waste	Generation of industrial, municipal waste	Accumulation to date	Expenditure on collection and treatment, recycling rates
5. Natural disaster			

Source: Hammond et al, 1995

Figure 15 *Matrix of Environmental Indicators*

The Unseen Perils of Technological Progress: The Case of CFCs

CFCs represent one of the basic conundrums of modern technological advance. Invented in the 1930s, CFCs were a major technological advance that promoted human welfare. They are colourless, odourless, non-corrosive gases that are cheap to produce and provide many advantages. They are excellent refrigerants, heralding the advent of capacities to preserve foods and render exceptionally hot environments comfortable for human habitation. They are also good insulators, protecting people and goods against excessive heat and cold, thus also rendering the natural environment more comfortable for human habitation. The comforts of modern life are now unthinkable without refrigeration, air conditioning and insulation. Moreover, when invented, it was found that the gases simply went away, doing no harm to any known sink. Forty years later it was theorized that these gases migrated slowly to the stratosphere and produced chemical reactions destroying the ozone layer that protects the earth against the harmful effects of ultraviolet radiation. Slightly more than a decade after this theorizing, evidence was sufficient to prove that the gases did indeed have disastrous ozone destroying capacities. By the end of the 1980s, the world community had come together to produce a global agreement to phase out the production of these gases and to replace them with less harmful ones.

It is most likely that any major technological advances made today will soon be found to have more deleterious effects than originally thought. There is no panacea, and no return to Eden. Whatever progress we make as a species will undoubtedly be found to have costs and require future adjustments.

raising the earth's temperature through the greenhouse effect have been combined into their CO_2 equivalent. The gases are carbon dioxide, methane, nitrous oxide, chlorofluoro-carbons (CFCs) and halons. Since the gases have different atmospheric lifetimes, and different rates of earth-emitted radiation absorption, they have been weighted by these factors and combined into a Global Warming Potential (GWP). These were calculated for a ten-year period, 1981 to 1991, and showed a 16 percent decline. The goal of the Dutch government is to reduce these emissions by another 15 percentage points by the year 2000.

CFC Emissions

CFCs and halons rise to the stratosphere and deplete the ozone layer, which protects the earth from harmful ultraviolet radiation. Used as refrigerants, propellants and insulants, these gases have been closely associated with much of what is considered progress in the development of modern urban industrial society. In the past two decades their ozone destroying capacities have been documented and an international agreement has been crafted to phase out their production by the end of this century. Like greenhouse gases, the ozone depleting capacity of these gases depends on their lifetime in the atmosphere and their chemical capacities to destroy ozone, both of which vary for different gases. The Dutch combined these into an ozone depleting index, measured the emissions over the same ten-year period, and found a 56 percent decline. The goal, however, was to reduce these emissions to zero by 1995.

Acidic Emissions

Many air pollutants, especially from fossil fuel consumption, add acids to the atmosphere, which can harm buildings, and destroy trees

CO_2 is emitted by fossil fuel combustion, but it is also absorbed by plants in the photosynthesis process. Thus some level of carbon emission is sustainable. For some of the pollutants, the Dutch exercise estimated a sustainable level, but it was not done for all. Reviewing this exercise briefly will provide an illustration of what can be done, but also of the problems that remain to be resolved.

Greenhouse Gas Emissions

Five trace gases that play a major role in

and fish as they fall to earth in rain. They can also increase the acidity of soil, reducing its life-supporting capacity. The Dutch measured the acid deposits in the soil and expressed them in acidification equivalents per hectare per year. These, too, showed a substantial decline (39 percent) over the past ten years, and the goal is to reduce emissions by another 46 percentage points by the year 2010. For this type of pollution the Dutch could estimate a sustainable level of acidification. Although the goal of 2010 is an ambitious one, it still implies a level that is considered sustainable in the long run.

Problem

In this case, by measuring acid depositions in the soil, the Dutch could see the impact of both domestic and foreign sources of emissions. While they can develop national policy responses to their own emissions, however, they will have little direct control over those of other countries, and must support international conventions to deal with this problem.

Nutrient Emissions

Plant nutrients, like phosphates and nitrogen, which are released in chemical fertilizers, increase algae in water bodies, leading to a shortage of oxygen that chokes off other forms of life. The process is known as eutrophication. The Dutch exercise measure nutrient emissions, expressed in eutrophication equivalents and tracked these over the decade. Here they found only a 10 percent decline, against a goal of another 71 percentage point drop by the year 2000. This level is considered sustainable, but the difficulty of reaching the goal is clear, especially for such a major agricultural producer.

Problem

There is another problem with this particular measure that beguiles the entire global community. Emissions do not honour national boundaries. In this case the Dutch measured only domestic emissions, thus ignoring the heavy load deposited by upstream polluters and carried to Holland by the rivers that debouch into the North Sea through the Netherlands.[31]

Toxic Emissions

Modern industrial society produces many toxic wastes. Radioactive wastes and heavy metals pose the most serious risks. Toxic chemicals produced or used in industrial processes have become a major problem, and increasingly subject to control. In agriculture poisonous pesticides are also extensively used. The Dutch combined all of these into one index, a hazardous pollution equivalent, weighting the physical amounts by their toxicity and lifetimes. In the past ten years the index rose to a peak in 1985, then began a more precipitous decline, bringing an 11 percent decline in the decade. The goal is to reduce emissions by another 29 percent by the end of this decade, to a level that is considered sustainable.

Problem

And yet another problem is illustrated by this measure. For pesticides the Dutch distinguished between agricultural and non-agricultural emissions and only considered the latter. In effect, agriculture was considered too important an industry to be saddled with the demand to reduce pesticide use. It is thus unclear whether the estimate of sustainable levels is truly sustainable. In other agricultural systems the use of pesticides has been found to increase rather than to decrease insect damage, leading

31 The Dutch also suffer another form of pollution by being at the end of a watershed that drains many European countries. Tilling agricultural fields to increase run-off and thus food production, and straightening river courses to facilitate water transportation along the Rhine results in extensive flooding in the Netherlands. The country has little control over these upstream changes, and remedying the situation will require difficult international agreements.

to the development of new systems of Integrated Pest Management (IPM), which greatly reduce poisonous emissions.

Solid Waste Emissions

Solid wastes now represent a growing problem for all modern governments. But they represent a major opportunity as well, since they can be turned into valuable resources. Solid wastes can be buried and mined for combustible methane gases, or incinerated, recycled and reused, which is now a fast growing industry world wide. In this exercise the Dutch simply measured the gross weight of solid wastes dumped, expressed in millions of tons of waste equivalents per year. Again, calculation over the ten year period showed only a modest decline of about 8 percent, with a goal of reducing by another 58 percentage points, down to just one third of the 1981 level. No level of sustainable dumping was estimated.

These five indices were then combined in a composite pollution index, based on the gap between the current value of the indicator and the long term policy goal of estimated sustainability. The composite index showed a decline of about 15 percent in the decade. There is progress, but also a long way to go.

This exercise illustrates a useful strategy for tracking environmental pollution over time. The calculations were made here for the entire nation, but a national planning activity could also estimate emissions for smaller administrative units, and relate these to population dynamics.

Resource Depletion

Natural resources are depleted by human activity. Non-renewable resources like oil and minerals are mined from the earth. Renewable resources like forests are cut down for timber, fuelwood and for transformation to agricultural

use. Water is pumped from underground aquifers and soil is depleted through erosion, waterlogging, salinization or compacting. Renewable resources can be restored by investments in such things as reforestation, water harvesting and soil protection. An aggregate resource depletion index has been calculated by constructing a ratio of resource use to capital formation. There is now active experimentation in developing such indicators, relying on standard national income accounting and adjustments made for environmental valuing. Such indices can be done in aggregate for a nation or its sub-national administrative units. It can also be done for sectors and specific resources.

There is not yet a standard, cross-nationally comparable set of accounts that would permit estimates of resource depletion for all countries, but there is movement toward such accounts. At present we must be content with the idea that resource depletion indices can be calculated, and related to any of our population indicators. In the absence of such aggregate indicators, one can still count areas of forests depleted or degraded, amount of land degraded, and excesses of water use over recharge capacities.

Ecosystem Risk and Biodiversity Protection

If we have not yet developed good aggregate indicators for resource depletion, we are even further behind in measuring biodiversity. We have lists of endangered species and of 'protected areas'. Species lists are useful to indicate areas where action may be required, but the actions to preserve species are not well developed. Conventions in trade of endangered species have been developed, but violations are extensive, and the conventions are controversial when they destroy the livelihood of certain groups of hunters. Lists of protected areas have been produced for some years, but they are especially deficient in that they seldom contain information on the extent to which human use intrudes illegally into areas only designated as protected. It is estimated, for example, that 85 percent of all protected areas in Latin America contain local populations living in and using the resources of the areas (McNeely, 1994).

There are also organized efforts to collect and preserve specific plant species and varieties in large scale germ banks. While these represent good efforts, they cover only a small portion of plants. Moreover, in these cases we confront biodiversity at the genetic or species level. Its is now widely recognized, however, that we must protect biodiversity at the ecosystem level. Unfortunately, there is as yet no widespread agreement on how this should be done. We have no aggregate indices of ecosystem health. Moreover, an ecosystem's ability to support biodiversity is highly location specific, making it difficult to construct national aggregate indices.

There are, however, effective means to monitor terrestrial ecosystems through digital mapping. This is a fast growing technology, greatly enhanced by the advances satellite technology and sophisticated GIS procedures. Digital mapping and GIS make it easy to create map overlays to record the spatial distribution of human activities and habitat change. Since these data are computerized, it is possible to develop simulations of future scenarios under different assumptions of human activity or ecosystem change. These can be combined with measures of ecosystem sensitivity, indicating the degree of risk that an ecosystem would not be able to support a given level of biodiversity. Thresholds for biodiversity support can be established, ecosystem risk indicators developed, and targets can be set in conjunction with the Biodiversity Convention.

FAO is currently developing map-based indicators of ecosystem risk to help in agricultural planning (FAO, 1993). A major global change group, the Consortium for International Earth Science Information Network (CIESIN) is making available digital maps showing population concentrations and distributions for many countries of the world. The WRI is spearheading a project to prepare ecosystem risk maps for a series of African countries (WRI, 1994). This is thus an area in which there is rapid growth. Despite the lack of well developed and commonly accepted aggregate indicators, the digital mapping technology make it possible for government planners to examine human activities and ecosystem change for the purposes of establishing effective policies.

Human Impact Exposure Indicators

We have touched on most of the issues here in discussions above on population, urbanization and density. Human health can be effectively assessed for many less developed countries by such indicators as infant or maternal mortality rates. These outcomes can be related to a variety of indicators of environmental and social conditions that produce the outcome: water and air quality, sewage disposal, housing, poverty and nutrition. As the matrix shows each of these pressure and state measurements can be related to policy responses that are well known to have an impact on the condition.

Summary Indicators

While work is progressing rapidly on the development of aggregate indicators of environmental impact, it may be useful and necessary to begin working with a few readily available indicators of environmental well being. We can propose four simple measures here, of changes for which data usual exist or can be readily obtained, and which may provide crude indicators of the major types of human impact.

Air quality

This is now commonly measured especially in urban areas, where a variety of important gasses are measured, including nitrous oxides, ozone, and acid levels. In addition it is common to measure particulate matter. There are useful standards for all of these, above which levels of human health may be affected, or at which environmental degradation is signalled.

Nutrients

Nutrients, or nitrogen and phosphate, often in the form of fertilizers and washing products, are discharged and have a destructive impact on lakes and streams, increasing algae levels and choking off other forms of life. Estimates can often be made from sales or distribution of fertilizer. It is also possible to organize school groups to measure water quality for indicators of this kind of impact.

Water quality

The simplest measures are those of human wastes in water sources. These, too, can often be monitored effectively by mobilizing school classes, where children learn about environmental degradation while they are actually taking the measures.

Resource depletion

Can more readily be assessed by deforestation, which can be done from mapping, or now quite easily from satellite imagery. There are also extensive estimates being made of deforestation by both government agencies and non-governmental environmental organizations. In addition, it is useful, but more difficult, to obtain estimates of soil erosion and of the degradation of agricultural land through salinization or waterlogging.

If measures are made for these four conditions, and if data collection were organized with relatively simple techniques that could be widely applied, planners would have useful indicators to place alongside the population data both to learn more about population–environment dynamics, and to develop effective interventions that can mitigate harmful human impacts and work toward protecting the wellbeing of both people and the ecosystem.

Conclusion

Here, then, is a great variety of population and environmental indicators that can be used to assess the dynamic relationships between population and the environment. Some are widely available, and though simple aggregate measures, like IMR, are sensitive indicators of the level of human and ecosystem health. There is also a very rapid development of both indicators and processes for linking these to policy decisions that can lead to effective interventions or responses. The next problem is to determine how best to put the indicators together for national level planning. To address these issues, we need to be concerned with two things:

- the business of organizing efforts for producing national strategies for sustainable development, and
- the frameworks or models that actually link the different indicators together.

These issues are discussed in detail in Parts II and III above.

Bibliography

Abernathy, Virginia, 1993, *Population Politics: The Choices that Shape our Future*, (New York: Plenum Publishing Co).

Agbo, Valintin, Nestor Skopon, John Hough, and Patrick C West, 1993, 'Population–Environment Dynamics in a Constrained Ecosystem in Northern Benin', in Ness, Drake and Brechin, 1993 op cit.

Ashford, Lori, 1995, 'New Perspectives on Population: Lessons from Cairo,' *Population Bulletin*, 50 (1), March, (Washington, DC: The Population Reference Bureau).

Australian Academy of Sciences, 1993, *Population and Development in the Asia–Pacific Region*, (Canberra: Australian Academy of Sciences).

Barton, Tom, Grazia Borrini-Feyerabend, Alex de Sherbinin and Patrizio Warren, 1996, *Our People, Our Resources: Supporting Rural Communities in Participatory Action Research on Population Dynamics and the Local Environment*, (Gland, Switzerland: IUCN).

Bongaarts, Jon, 1988, 'The supply demand framework for the determinants of fertility: an alternative implementation', *Population Studies*, 47 (3) pp 437–456.

Boserup, Esther, 1965, *The Conditions of Agricultural Progress: The Economics of Agrarian Change under Population Pressure*, (Chicago: Aldine).

—, 1981, *Population and Technology*, (Oxford: Basil Blackwell).

Braat, L C, 1991, 'Sustainable Development and Ecological Modelling,' in Gilbert and Braat, eds, op cit, pp 169–190.

Brown and Wycoff-Baird, *Designing Integrated Conservation Development Projects,* referred to in USAID 1994.

Caldwell, John, 1986, 'Routes to Low Fertility in Poor Countries,' *Population Development Review*, 12, (2), pp 171–220.

—, 1976, 'Towards a Restatement of Demographic Transition Theory,' *Population and Development Review*, 2–3/4 (September–December) pp 321–66.

—, 1994, 'The Course and Causes of Fertility Decline,' IUSSP Distinguished lecture series of population and development, (Liege, Belgium: IUSSP).

Carew-Reid, J, Prescott-Allen, R, Bass, S, and Dalal-Clayton, D B, 1994, *Strategies for National Sustainable Development: A Handbook for their Planning and Implementation*, (London: Earthscan Publications).

Carson, Rachael, 1964, *Silent Spring*, (Boston: Houghton Mifflin).

Clelland, John, and Chris Scott, 1987, *The World Fertility Survey*, (Oxford: Oxford University Press).

Clark, William, 1992, 'The Role of Population Growth in Environmental Degradation, Towards a Comparative Historical Analysis of Country Level Data', manuscript, (Cambridge, MA: Harvard University John F Kennedy School of Government).

Coleman, D A, 1992, 'The World on the Move? International Migration in 1992,' paper for the European Population Conference, Geneva, Switzerland, March 1993, E CONF 84RM EUR WP 1, sponsored by UNECE, Council of Europe, UNFPA.

Daly, Herman and John B Cobb, 1989, *The Common Good* (Boston, MA: Beacon Press).

—, 1992, 'Allocation, Distribution and Scale: Towards and Economics that is Efficient, Just, and Sustainable,' *Ecological Economics*, December.

—, 1994, 'Fostering environmentally sustainable development: four parting suggestions for the World Bank,' *Ecological Economics*, 10, pp 183–87.

Davis, Kingsley, 1967, 'Population Policy, Will Present Programs Succeed', Science, 158, 3082 (November), pp 730–39.

de Sherbinin, Alex, 1993, 'Population and Consumption Issues for Environmentalists,' a literature search prepared for the Pew Charitable Trusts' Global Stewardship Initiative, (Washington, DC: Population Reference Bureau).

Drake, Michael, ed, *Population in Industrialization*, (London: Methuen & Co Ltd).

Dyson, Tim, 1994, 'Population Growth and Global Food Production: Recent Global and Regional Trends,' *Population Development Review*, 20 (2), June.

Ehrlich, Paul, 1994, 'Population and Sustainable Development,' in *Environmental Awareness*, 17 (2).

Ehrlich, Paul and Ann Ehrlich, 1972 and 1990, *The Population Bomb*, (New York: Addison Wesley)

—, 1991, *Healing the Planet*, (New York: Addison Wesley).

—, and John Holdren, 1971, 'Impact of Population Growth,' *Science*, 171, 26 March, pp 1212–1217.

ESCAP, 1975, *'Study on Family Planning Administration in the Republic of Korea: Interpretive Summary'*, (Bangkok: ESCAP Population Division).

FAO and UNESCO, 1985, *Carrying Capacity Assessment: with a pilot study of Kenya*, Report W/R8012, (Rome: FAO, and Paris: UNESCO).

FAO, 1986, *Potential Population Supporting Capacities of Lands in the Developing World*, (Rome: FAO).

FAO, together with the International Institute for Applied Systems Analysis (IIASA) 1993, *Agroecological Assessment for National Planning: The Example of Kenya*, FAO Soils Bulletin (67, (Rome: FAO).

Feldstein, Hillary Sims, and Susan V Poats, eds, 1989, *Working Together: Gender Analysis in Agriculture*, (West Hartford, CT: Kemurian Press).

Finkle, Jason S, and Gaykl D Ness, 1985, *Building Effective Delivery Systems*, Report to USAID, Office of Population, (Ann Arbor, MI: Center for Population Planning).

Futures Groups, The, nd, *RAPID IV: Resources for the Awareness of Population Impacts on Development,* (Washington, DC, The Futures Group).

Gilbert, A J and L C Braat, eds, 1991, *Modelling for Population and Sustainable Development*, (London: Routledge).

Grainger, Alan, 1993, 'Population as a Concept and Parameter in the Modelling of Deforestation,' in Ness, Drake and Brechin, op cit.

Grosse, Scott, 1993, *Schistosomiasis and Water Resource Development: A re-evaluation of an Important Environment-Health Link*, (Madison, WI: The Environmental and Natural Resources Policy and Training Project, EPAT).

Hammond, Allen, Albert Adriaanse, Eric Rodenburg, Dirk Bryant, and Richard Woodward, 1995, *Environmental Indicators: A Systematic Approach to Measuring and Reporting on Environmental Policy Performance in the Context of Sustainable Development*, (Washington, DC: World Resources Institute).

Harrison, Paul, 1992, *The Third Revolution: Environment, Population and a Sustainable World*, (London: I B Taurus & Co).

Haub, Carl, 1987, 'Understanding Population Projections,' *Population Bulletin*, 42, (4), December (Washington, DC: Population Reference Bureau).

Hernandez, Donald, 1984, *Success or Failure: Family Planning Programs in the Third World*, (Westport, CT: Grenwood Press).

ILO, 1992, *Migration*, (Geneva: International Labour Organization).

IPCC (Intergovernmental Panel on Climate Change) 1995, *Second Assessment Report*, Working Group III.

IUCN, 1984, *Traditional Life-styles, Conservation and Rural Development*, Commission on Ecology Papers Number 7, (Gland: Switzerland: IUCN).

—, 1987, *Population and Sustainable Development,* Task Force report, (Gland: IUCN).

—, 1991, *Caring for the earth: A Strategy for Sustainable Living*, with UNEP and WWF, (London: Earthscan).

—, 1993, *Women in Conservation: Tools for Action and a Framework for Analysis*, by Dounia Loudiyi and Alison Meares, (Washington, DC: IUCN).

—, 1996, *Indigenous Conservation in the Modern World: Case Studies in resource exploitation, Traditional Practice, and Sustainable Development,* Inter-Commission Task Force on Indigenous Peoples, (Gland, Switzerland: IUCN).

IWGIA, 1992, *Yearbook, 1991*, (Copenhagen: International Working Group on Indigenous Peoples).

Jacobson, Jodi, 1988, 'Environmental Refugees: A Yardstick of Habitability,' *Worldwatch Paper 86*, (Washington, DC: Worldwatch Institute).

—, 1994, 'Abortion and the Global Crisis in Women's Health,' in Mazur, ed, 1994 op cit.

Kallenberg, John and Herman Daly, 1993, 'Counting User Cost in Evaluating Projects involving Depletion of Natural Capital: World Bank Best Price and Beyond', World Bank Environment Department, October.

Kane, Hal, 1995, *The Hour of Departure: Forces that Create Refugees and Migrants*, Worldwatch Paper 125, (Washington, DC: Worldwatch Institute).

Kelley, Allen C, 1991, 'The Human Development Index: "Handle with Care"', *Population Development Review*, 17 (2), (June).

King, J, 1987, *Beyond Economic Choice*, (Edinburgh: University of Edinburgh, and Paris: UNESCO)

—— and M Slesser, 1988, 'Resource accounting for development planning,' *World Development*, April.

King, J, 1991, 'The ECCO Approach to planning sustainable development,' in Gilbert and Braat, eds, op cit, pp 18–35.

Livi-Bacci, Massimo, 1989, *A Concise History of World Population*, (Cambridge, MA: Blackwell).

Lutz, Wolfgang, ed, 1994, *Population Development Environment: Understanding their Interactions in Mauritius*, (Berlin: Springer-Verlag).

MacKellar, F Landis, Anne Goujon, Wolfgang Lutz and Christopher Prinz, 1995 'Population, Number of Households, and Global Warming,' *POPNET*, Population Network Newsletter (27), Fall, pp 1–3.

Mamdani, Mahmood, 1972, *The Myth of Population Control; Family, caste, and class in an Indian village*, (New York, Monthly Review Press).

Marcoux, Alain, 1994, *Population and Water Resources*, manuscript, Population and the Environment: a review of isssues and concepts for population programme staff, (Rome: FAO)

Martine, George, 1995 'Brazil's Remarkable Fertility Decline: A fresh look at key factors,' Working Paper Series 95:04 (Cambridge, MA: Harvard Center for Population and Development Studies).

Mazur, Laurie A, 1994, *Beyond the Numbers: A reader on Population, Consumption and the Environment*, (Washington, DC: Island Press).

McFalls, Joseph A Jr, 1991, 'Population: A Lively Introduction,' *Population Bulletin*, 46, 2, October, (Washington, DC: Population Reference Bureau).

McNeely, Jeffrey, J Harrison, and P Dingwall, eds, 1994, *Protecting Nature: Regional Review of Protected Areas, IVth World Congress on National Parks and Protected Areas, Caracas, Venezuela*, (Gland: IUCN).

—, 1995, 'Cities, Nature and Protected Areas: A General Introduction,' II Symposium on Natural Areas in Conurbations and on City Outskirts, Barcelona, 25–27 October 1995, (Gland: IUCN).

McNeely, Jeffrey and Gayl Ness, 1995, 'People, Parks and Biodiversity,' in *Human Population, Biodiversity and Protected Areas: Science and Policy Issues*, (Washington, DC: American Academy for the Advancement of Science).

Meadows, Donella H, Dennis L Meadows and Jorgen Randers, 1972, *The Limits to Growth*, (London: Pan Books Ltd).

—, 1992, *Beyond the Limits: Confronting Global Collapse, Envisioning a Sustainable Future*, (London: Earthscan).

Morris, Morris David, 1977, *The PQLI: Measuring Progress in Meeting Human Needs*, (Washington, DC: Overseas Development Council).

Mosher, Caroline O N, 1993, *Gender Planning and Development: Theory Practice and Training*, (London: Routledge).

National Research Council, 1994, *Assigning Economic Value to Natural Resources*, (Washington, DC: National Research Council).

Ness, Gayl D, 1964, *Bureaucracy and Rural Development in Malaysia*, (Berkeley: University of California Press).

—, and Hirofumi Ando, 1984, *The Land is Shrinking: Population Planning in Asia*, (Baltimore: Johns Hopkins University Press).

—, William D Drake, and Steven R Brechin, eds, 1993, *Population Environment Dynamics: Ideas and Observations*, (Ann Arbor: University of Michigan Press).

—, 1994, *Population and Environment: Frameworks for Analysis*, (Madison, WI: The Environmental and Natural Resources Policy and Training Project: EPAT).

Notestein, Frank W, 1945, 'Population, the Long View,' in Theodore W Schultz, ed, *Food for the World*, (Chicago: University of Chicago Press).

OAS, 1990, *Disaster Planning, and Development: Managing Natural Hazards to Reduce Loss*, (Washington, DC: Organization of American States).

OECD, 1993, *Agricultural and Environmental Policy Integration: Recent Progress and New Directions*, (Paris: OECD).

OECD, 1994, *Environmental Indicators*, (Paris: OECD).

Opschoor, J B, 1991, 'Economic Modelling and sustainable development,' in Gilbert and Braat, eds, op cit, pp 191–209.

Pagiola, Stefano, 1995, 'Environmental and Natural Resource Degradation in Intensive Agriculture in Bangladesh,' *Environmental Economics Series*, Environment Department Papers (15, (Washington, DC: The World Bank).

Picardi, Anthony, 1974, 'A Systems Analysis of Pastoralism in the West African Sahel,' Annex to *A Framework for Evaluating Long Term Strategies for development of the Sudan Sahel Region*, Report CPA 74–9, MIT, (Cambridge, MA: Center for Policy Alternatives).

Phillips, James E and John A Ross, eds, 1992, *Family Planning Programmes and Fertility*, (Oxford: The Clarendon Press).

Population Action International, 1993, *Challenging the Planet: Connections Between Population and the Environment*, (Washington, DC: Population Action International).

Prichett, Lant, 1994, 'Desired fertility and the impact of population policies,' *Population and Development Review*, 20(1) March, pp 1–56 (New York: The Population Council).

PAI, 1994, *Global Migration: People in the Move,* wall chart with map and figures, (Washington, DC: Population Action International).

Rabinovitch, Jonas, and John Hoehn, 1995, 'A Sustainable Urban Transport System: The "Surface Metro" in Curtiaba, Brazil', *EPAT Working Paper* (19), (Madison, Wisconsin: The Environmental and Natural Resources Policy and Training Project).

Rathberger, E and B Kettel, eds, 1988, *Women's Role in Natural Resource Management*, A collection of papers presented at a meeting on African women's roles in natural resource management, (Ottawa: International Development Research Center).

Repetto, Robert, 1987, *Population, Resources, Environment: An Uncertain Future*, Population Bulletin 42 (2 (Washington, DC: Population Reference Bureau).

—, 1989, *Wasting Assets: Natural Resources in the National Income Accounts*, (Washington, DC: The World Resources Institute).

—, 1994, *The 'Second India' Revisited: Population, Poverty, and Environmental Stress over Two Decades*, (Washington, DC: World Resources Institute).

Robinson, Nicholas A, ed, 1993, *Agenda 21: Earth's Action Plan*, (New York: Oceana Publications).

Ross, John A, W Parker Mauldin, Vincent C Miller, 1993, *Family Planning and Population: A Compendium of International Statistics*, (New York: The Population Council).

—, and Elizabeth Frankenberg, 1993, *Findings from Two Decades of Family Planning Research*, (New York: The Population Council).

RPG, 1992, *Migration and the Environment,* report on a conference convened in Nyon, Switzerland, June 1992, jointly sponsored by the Swiss Foreign Ministry, the International Organization for Migration and the Refugee Policy Group, (New York: Refugee Policy Group, and Geneva: International Organization for Migration).

Sadik, Nafis, 1991, *Population Policies and Programmes: Lessons Learned from Two Decades of Experience*, (New York: UNFPA).

Sanderson, Warren G, 1994, 'Models of Demographic, Economic and Environmental Interactions,' in Lutz, 1994, op cit, pp 33–71.

Slesser, M, 1987, 'Net energy as an energy planning tool', *Energy Policy*, June, pp 228–36.

Smil, Vaclav, 1994, 'How Many People Can the Earth Feed?', *Population Development Review*, 20, (2), (June) pp 255–92.

Smith, Kirk R, 1993, *The Most Important Chart in the World*, UN University Lectures 6, (Tokyo: The United Nations University).

Stephenson, Patricia, Marsden Wagner, Mihaela Badea, and Florina Servanescu, 1992, 'Commentary: The Public Health Consequences of Restriucted Induced Abortion – Lessons from Romania ' *American Journal of Public Health*, 82/10, pp 1321–1331.

Stren, Richard, Rodney White and Joseph Whitney, eds, 1992, *Sustainable Cities: Urbanization and the Environment in International Perspective*, (Boulder, CO: Westview Press).

UN, 1992, *World Urbanization Prospects, The 1992 Revision*, (New York: United Nations Population Division).

—, 1994, *World Population Prospects, The 1994 Revision*, (New York: United Nations Population Division).

—, 1994A, *World Urbanization Prospects, The 1994 Revision*, (New York: United Nations Population Division).

—, 1994B, *Population and the Environment in Developing Countries: Literature Survey and Research Bibliography*, Draft, (New York: United Nations).

—, 1995, *Population and Development: Programme of Action Adopted at the International Conference on Population and Development*, Cairo, 5–13 September 1994, STA/ESA/SER A/149, (New York: United Nations).

—, 1995A, *Implementing Sustainable Development: Experiences in Sustainable Development Administration*, DDSMS/SEM 95/1 INT-91-R71, (New York: United Nations Department for Development Support and Management Services).

—, 1995B, *The Copenhagen Declaration and Programme of Action*, World Summit for Social Development, 6–12 March, (New York: United Nations).

—, 1995C, *Global Population Data Base*, (New York: United Nations Population Division).

UN Statistical Office, 1993, *Integrated Environmental and Economic Accounting*, (New York: United Nations).

UNDP, 1993, *Human Development Report, 1993*, (New York: United Nations Development Program).

UNGA 1994, *Inauguration of the International Year of the World's Indigenous People*, Centre for Human Rights, MHR/93/14 GE 93-15220, (New York: United Nations General Assembly).

UNICEF, 1994, *The State of the World's Children 1994*, (New York: UNICEF).

UNFPA, 1994, *The State of World Population: Choices and Responsibilities*, (New York: UNFPA).

—, 1995 *The State of World Population: Decisions for Development*, (New York: UNFPA)

USAID, Environmental and Natural Resources Information Centre, 1994, *Conservation of Tropical Forests and Biodiversity: Program Update*, Initial Working Document, (Washington DC: USAID).

Vitusek, Peter, Paul R Ehrlich, and Pam A Matson, 'Human Appropriation of the Products of Photosynthesis,' *Bioscience*, 36(3): 368–373, 1986.

Vosti, Steven A, and Julie Witcover, 1995, 'A 'Decomposition' of Population and Environmental Linkages: Towards a Conceptual Framework,' manuscript, presented at the Population Association of America (PAA) annual meetings, April 6–8, 1995, San Francisco, California.

Whyte, Martin, K 1978 *The Status of Women in Preindustrial Societies*, (Princeton: Princeton University Press).

Wils, Anna, 1995, *PDE-Cape Verde, A systems study of population, environment and development* Manuscript, (Laxenburg, Austria: IIASA).

World Bank, 1992, *World Development Report*, (Washington, DC: The World Bank).

—, 1993, *Effective Family Planning Programs*, (Washington, DC: The World Bank).

WRI, 1994, *World Resources, 1994–95*, (Washington, DC: World Resources Institute).

—, with the World Conservation Monitoring Centre, *African Data Sampler: An Internationally Comparable Map and Georeferenced Data Base* (Washington, DC: WRI).

Zhang, Qun, 1994, *The Impact of International Population Assistance*, Ph D Dissertation, School of Public Health, (University of Michigan: Ann Arbor, MI).

Glossary

BOD	biological oxygen demand
CPR	contraceptive prevalence rate
DEP	development–environment–population
ECAFE	Economic Commission for Asia and the Far East
EPZ	Export Promotion Zone
ESCAP	United Nations Economic and Social Commission for Asia and the Pacific
GAD	gender and development
GIS	geographical information system
GNP	Gross National Product
GWP	Global Warming Potential
HDI	Human Development Index
ICDP	USAID's Integrated Conservation and Development Projects
ICPD	International Conference on Population and Development
IIASA	International Institute of Applied Systems Analysis
IMR	Infant Mortality Rate
IPCC	Intergovernmental Panel on Climate Change
IPM	Integrated Pest Management
IPPF	International Planned Parenthood Federation
IUD	intra-uterine device
IUSSP	International Union for the Scientific Study of Population
NGO(s)	non-governmental organization(s)
NPER	national population–environment review
OECD	Organization for Economic Cooperation and Development
PAA	Population Association of America
PDE	IIASA's population–environment–development model
PEC	primary environmental care
PEN	population–environment network
PPP	parity purchasing power
PQLI	Physical Quality of Life Index
PRA	participatory rapid appraisal
QOL	quality of life
SDDR	Socio-Demographic Dependency Ratio
SEEA	System for Environmental and Economic Accounting
SNSD	Strategies for National Sustainable Development
TFAP	Tropical Forest Plan of Action
TFR	Total Fertility Rate
TVA	Tennessee Valley Authority
UN	United Nations
UNCED	United Nations Conference on Environment and Development
UNDP	United Nations Development Programme
UNEP	United Nations Environment Programme
UNFPA	United Nations Fund for Population Activities
WID	Women in Development
WRI	World Resources Institute

Index